Our Amazing Planet

TIME® LIFE BOOKS

Time-Life Books is a division of Time Life Inc.
Time-Life is a trademark of Time Warner Inc.
and affiliated companies.

Conceived and produced by Weldon Owen Pty Limited
59 Victoria Street, McMahons Point, NSW, 2060, Australia
A member of the Weldon Owen Group of Companies
Sydney • San Francisco • London
© 2001 Weldon Owen Inc.

TIME LIFE INC.
Chairman and Chief Executive Officer: Jim Nelson
President and Chief Operating Officer: Steven Janas
Senior Executive Vice President and Chief Operations Officer: Mary Davis Holt
Senior Vice President and Chief Financial Officer: Christopher Hearing

TIME-LIFE BOOKS
President: Larry Jellen
Senior Vice President, New Markets: Bridget Boel
Vice President, Home and Hearth Markets: Nicholas M. DiMarco
Vice President, Content Development: Jennifer L. Pearce

TIME-LIFE TRADE PUBLISHING
Vice President and Publisher: Neil S. Levin
Senior Sales Director: Richard J. Vreeland
Director, Marketing and Publicity: Inger Forland
Director of Trade Sales: Dana Hobson
Director of Custom Publishing: John Lalor
Director of Rights and Licensing: Olga Vezeris

OUR AMAZING PLANET
Director of New Product Development: Carolyn M. Clark
Senior Editor: Robert Somerville
Director of Design: Tina Taylor
Project Manager: Jennifer L. Ward
Production Manager: Virginia Reardon

WELDON OWEN PUBLISHING
Chief Executive Officer: John Owen
President: Terry Newell
Publisher: Sheena Coupe
Associate Publisher: Lynn Humphries
Art Director: Sue Burk
Editorial Coordinator: Sarah Anderson
Production Manager: Helen Creeke
Production Assistant: Kylie Lawson
Vice President International Sales: Stuart Laurence

Managing Editor: Rosemary McDonald
Project Editors: Helen Bateman, Ann B. Bingaman, Helen Cooney, Jean Coppendale, Kathy Gerrard, Selena Quintrell
Text Editors: Ann B. Bingaman, Claire Craig, Emma Marks

Educational Consultants: Richard L. Needham, Deborah A. Powell
Designers: Sylvie Abecassis, Nicole Court, Kathy Gammon, Robyn Latimer, Michéle Lichtenberger, Giulietta Pellascio, Sue Rawkins, Melissa Wilton
Assistant Designers: Janet Marando, Kylie Mulquin
Visual Research Coordinators: Esther Beaton, Kathy Gerrard, Jenny Mills
Visual Researchers: Peter Barker, Karen Burgess, Annette Crueger, Carel Fillmer, Sue Liu, Amanda Parsonage, Amanda Weir

Text: Carson Creagh, David H. Levy, Linsay Knight, Sally Morgan

Illustrators: Mike Atkinson/Garden Studio; Graham Back; Kenn Backhaus; Andrew Beckett/Garden Studio; Greg Bridges; Lynette R. Cook, Simone End; Christer Eriksson; Nick Farmer/Brihton Illustration; Rod Ferring; John Francis; Sian Frances/Garden Studio; Mike Golding; Mike Gorman; Lorraine Hannay; Richard Hook/Bernard Thornton Artists, UK; Robert Hynes; Roger Kent/Garden Studio; Peter Kesteven/Garden Studio; David Kirshner; Frank Knight, Alex Lavroff; Mike Lamble; Jillian B. Luff; Iain McKellar; James McKinnon; Peter Mennim; Colin Newman/Garden Studio; Paul Newman; Kevin O'Donnel; Wendy de Paauw; Darren Pattenden/Garden Studio; Evert Ploeg; Marilyn Pride; Tony Pyrzakowski; Oliver Rennert; John Richards; Ken Rinkel; Andrew Robinson/Garden Studio; Trevor Ruth; Peter Schouten; Rod Scott; Steve Seymour/Bernard Thornton Artists, UK; Michael Saunders; Ray Sim; Kevin Stead; Steve Trevaskis; Rod Westblade; Simon Williams/Garden Studio

All rights reserved. No part of this book may be reproduced in any form or by any electronic or mechanical means, including information storage and retrieval devices or systems, without prior written permission from the publisher, except that brief passages may be quoted for reviews.

Color reproduction by Colourscan Co Pte Ltd
Printed by LeeFung-Asco Printers
Printed in China
10 9 8 7 6 5 4 3 2 1

School and library distribution by Time-Life Education,
P.O. Box 85026, Richmond, Virginia 23285-5026.

CIP data available upon request:
Librarian, Time-Life Books
2000 Duke Street
Alexandria, VA 22314

A Weldon Owen Production

Our Amazing Planet

CONSULTING EDITORS

David Ellyard
David H. Levy
Angela Milner
Eldridge M. Moores
Frank H. Talbot

Contents

DINOSAURS 6

THE ARRIVAL OF THE DINOSAURS
Before the Dinosaurs 8
What is a Dinosaur? 10
Dinosaur Hips 12
The Triassic World 14
The Jurassic World 16
The Cretaceous World 18

A PARADE OF DINOSAURS
Meat-eating Dinosaurs 20
Plant-eating Dinosaurs 22
Long-necked Dinosaurs 24
Armored, Plated & Horned Dinosaurs 26
Duckbilled Dinosaurs 28
Record-breaking Dinosaurs 30

UNCOVERING DINOSAUR CLUES
Fossilized Clues 32
Skeletons and Skulls 34
Footprints and Other Clues 36

LIFE AS A DINOSAUR
Raising a Family 38
Keeping Warm; Keeping Cool 40
Eating and Digesting 42
Attack and Defense 44

THE END OF THE DINOSAURS
Why Did They Vanish? 46
Surviving Relatives 48
Identification Parade 50

UNDER THE SEA 52

UNDERWATER GEOGRAPHY
Our Oceans 54
The Sea Floor 56
Sea Upheavals 58
Currents and Tides 60

LIFE IN THE SEA
The Seashore 62
Coastal Seas 64
Coral Reefs 66
Ocean Meadows 68
Life in the Twilight Zone 70
Life on the Ocean Floor 72

EXPLORING THE OCEANS
Submersibles 74
Research Ships 76

OCEAN MYSTERIES
Sea Legends 78
Where Did They Go? 80
Mysteries of Migration 82

EXPLOITING THE OCEANS
Oils and Minerals 84
The Perils of Pollution 86
Conserving the Oceans 88

VOLCANOES AND EARTHQUAKES 90

THE UNSTABLE EARTH
Fire Down Below 92
The Moving Continents 94
Ridges and Rift Valleys 96
Subduction 98
Hotspots 100

VOLCANOES
Volcanic Eruptions 102
Lava Flows 104
Gas and Ash 106

AFTER THE EVENT
　Craters and Calderas　108
　Volcanic Rocks and Landforms　110
FAMOUS VOLCANOES
　Mediterranean Eruptions　112
　Krakatau　114
　Iceland　116
　Mt. St. Helens　118
EARTHQUAKES
　Surviving an Earthquake　120
　Tsunamis and Floods　122
FAMOUS EARTHQUAKES
　The Great Kanto Earthquake　124
　Mexico City　126
　Californian Quakes　128

WEATHER　130

OUR WEATHER
　What is Weather?　132
　The Weather Engine　134
THE DAILY WEATHER
　Temperature and Humidity　136
　What are Clouds?　138
　Types of Cloud　140
　Thunder and Lightning　142
　Rain, Hail and Snow　144
　Fog, Frost and Ice　146
　Weather Wonders　148
WEATHER FORECASTING
　Weather Watch　150
　Forecasting　152
CLIMATE
　Winds and Currents　154
　World Climate　156
　Polar Zones　158
　Mountain Zones　160
　Temperate Zones　162
　Tropical Zones　164
　Desert Zones　166

CLIMATIC CHANGE
　Global Freezing　168
　Global Warming　170

BEYOND PLANET EARTH　172

OUR NEIGHBORHOOD
　The Solar System　174
　The Sun　176
　Mercury and Venus　178
　The Earth　180
　The Earth's Moon　182
　Mars　184
　Jupiter　186
　Saturn　188
　Uranus　190
　Neptune　192
　Pluto　194
　Comets　196
　Asteroids and Meteoroids　198
　Eclipses　200
OUR UNIVERSE
　The Universe　202
　The Life Cycle of a Star　204
　Strange Stars　206
　Galaxies　208
　The Milky Way　210
EXPLORING THE UNIVERSE
　Into Space　212
　Imagined Worlds　214
　Facts and Figures　216

　Glossary　218
　Index　222

Dinosaurs

- Why did duckbill dinosaurs hoot?
- Why are there so few dinosaur fossils?
- Which dinosaur whiplashed predators with a heavy club on its tail?

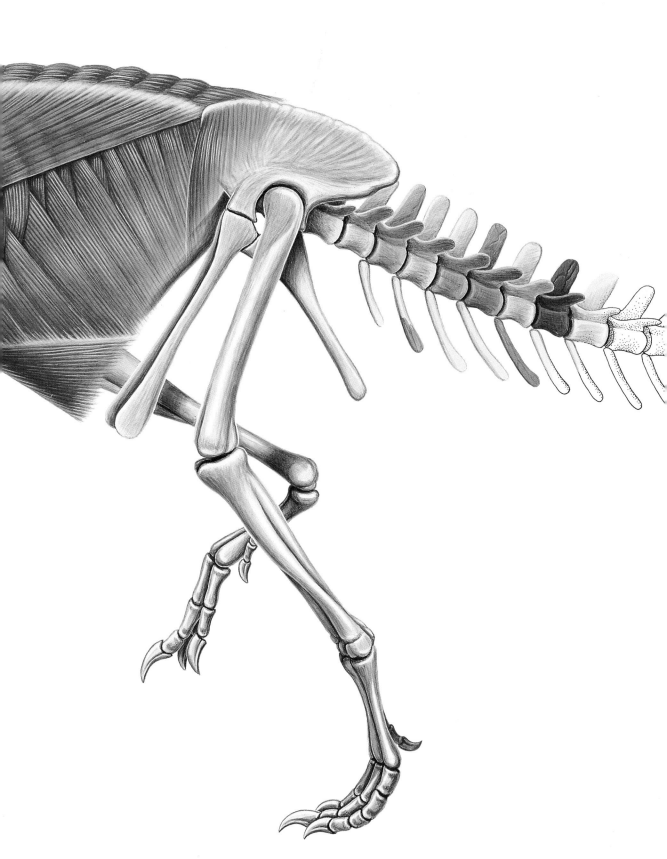

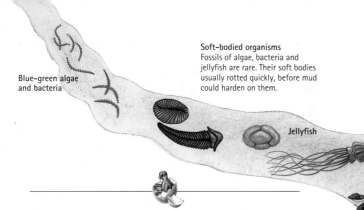

Blue-green algae and bacteria

Soft-bodied organisms
Fossils of algae, bacteria and jellyfish are rare. Their soft bodies usually rotted quickly, before mud could harden on them.

Animals with shells
Between 570 and 250 million years ago, there were more than 10,000 kinds of trilobites. They ranged from 1 in (3 cm) to 1½ ft (50 cm) in length.

Jellyfish

Ammonite

Trilobites

Pteraspis

Drepanaspis

Scorpion

• THE ARRIVAL OF THE DINOSAURS •

Before the Dinosaurs

Life on Earth began 4,600 million years ago. Its long history is divided into different periods, during which an amazing variety of life forms developed or died out. In the beginning, single-celled algae and bacteria formed, or evolved, in the warm seas that covered most of the planet.

In the Paleozoic Era, more complex plants and animals appeared in the sea: worms, jellyfish and hard-shelled mollusks swarmed in shallow waters and were eaten by bony fish. When plants and animals first appeared on land, they were eaten by amphibians that had evolved from fish with lungs and strong fins. Some amphibians then evolved into reptiles that did not lay their eggs in water. Early reptiles developed into turtles and tortoises, lizards, crocodiles, birds and the first dinosaurs. They dominated the world for millions of years.

The first jawless fish
Armor-plated fish such as *Pteraspis* and *Drepanaspis* did not have jaws. They sucked up food from the mud or fed on plankton.

A bony fish
Dunkleosteus, a giant of late Devonian seas, grew to 11 feet (3.4 m). It grasped its prey in sharp, bony dental plates because it did not have teeth.

An amphibian
Ichthyostega could not expand or contract its solid rib cage. This 3-ft (1-m) long amphibian had to use its mouth to push air into its lungs.

An early reptile
Hylonomus, a 8-in (20-cm) long reptile, is known only from fossils found in the remains of hollow tree trunks, where it may have become trapped while hunting insects.

A mammal-like reptile
Dimetrodon, which was 10 ft (3 m) long, may have angled its "sail" to catch sunlight so that it could warm up quickly in the morning.

Thecodontian archosaurs
Ornithosuchus looked like *Tyrannosaurus*, but it was not a dinosaur. It had five toes on each hind foot while *Tyrannosaurus* had only three.

THE DINOSAUR RACE

Dinosaurs, which appeared in the Triassic Period, were descended from crocodile-like reptiles, whose legs sprawled at right angles from their bodies. *Euparkeria*, from the early Triassic Period, had straighter legs and carried its body off the ground. The legs of *Lagosuchus*, from the middle Triassic Period, were tucked beneath its body, and it walked on its hind legs. By the late Triassic Period, the predator *Ornithosuchus* looked a little like a dinosaur, but the earliest known dinosaur was *Eoraptor*. It appeared 228 million years ago.

Ornithosuchus 13 ft (4 m)

Euparkeria 2 ft (60 cm)

Lagosuchus 1 ft (30 cm)

Eoraptor 4 ft (1.2 m)

• THE ARRIVAL OF THE DINOSAURS •

What is a Dinosaur?

Dinosaurs were very special reptiles. Some were the size of chickens; others may have been as long as jumbo jets. These creatures were the most successful animals that have ever lived on Earth. They dominated the world for nearly 150 million years and were found on every continent during the Mesozoic Era, which is divided into the Triassic, Jurassic and Cretaceous periods. Like reptiles today, dinosaurs had scaly skin and their eggs had shells. The earliest dinosaurs ate meat, while later plant-eating dinosaurs enjoyed the lush plant life around them. Dinosaurs are called "lizard-hipped" or "bird-hipped," depending on how their hip bones were arranged. They stood on either four legs or two and walked with straight legs tucked beneath their bodies. Dinosaurs are the only reptiles that have ever been able to do this.

JURASSIC HUNTERS
In a scene from the Jurassic Period in North America, 69-ft (21-m) long *Apatosaurus* munches on a cycad as a 7-ft (2-m) long *Ornitholestes* pounces on a salamander disturbed by the grazing giant.

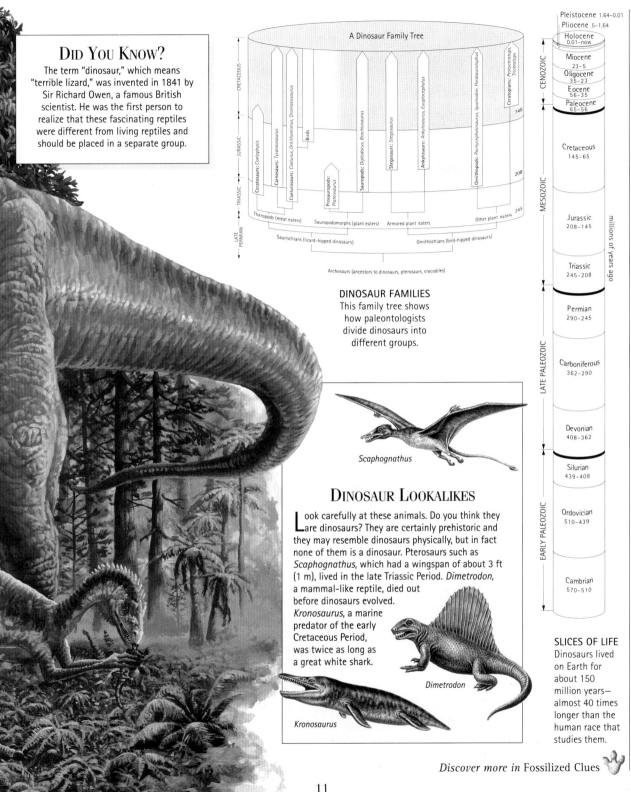

DID YOU KNOW?

The term "dinosaur," which means "terrible lizard," was invented in 1841 by Sir Richard Owen, a famous British scientist. He was the first person to realize that these fascinating reptiles were different from living reptiles and should be placed in a separate group.

DINOSAUR FAMILIES
This family tree shows how paleontologists divide dinosaurs into different groups.

DINOSAUR LOOKALIKES

Look carefully at these animals. Do you think they are dinosaurs? They are certainly prehistoric and they may resemble dinosaurs physically, but in fact none of them is a dinosaur. Pterosaurs such as *Scaphognathus*, which had a wingspan of about 3 ft (1 m), lived in the late Triassic Period. *Dimetrodon*, a mammal-like reptile, died out before dinosaurs evolved. *Kronosaurus*, a marine predator of the early Cretaceous Period, was twice as long as a great white shark.

Q: How are dinosaurs similar to living reptiles?

SLICES OF LIFE
Dinosaurs lived on Earth for about 150 million years—almost 40 times longer than the human race that studies them.

Discover more in Fossilized Clues

SPRAWLING
The ancestors of the dinosaurs sprawled on four legs, like a lizard. They had to use large amounts of energy to twist the whole body and lift each leg in turn.

HALFWAY UP
Some reptiles, such as today's crocodiles, have upright hind legs. As their bodies are off the ground, they can run on their hind legs for short distances.

ON TWO LEGS
A dinosaur's weight was supported easily by its straight legs, tucked under its body. As the body weight was balanced over the hips by the weight of the tail, some dinosaurs were bipedal (two-legged) and used their hands for grasping.

TWO-LEGGED PLANT EATER
This 43-ft (13-m) long *Edmontosaurus* has a typical ornithischian pelvis. The pubis points backwards and allows more space for the large intestines that plant eaters needed to digest their food. Conversely, in sauropods (plant-eating saurischians) the intestines are slung forward. This means the forward-pointing pubis does not get in the way of these four-legged (quadrupedal) dinosaurs.

• THE ARRIVAL OF THE DINOSAURS •

Dinosaur Hips

Dinosaurs walked upright with their legs beneath their bodies. No other reptiles have been able to do this. Dinosaurs had a right-angled joint at the top of the leg bone that fit into a hole in the hip bones. This allowed the limbs to be positioned under the body, so the weight of the dinosaur was supported, and all the joints worked as simple forward and backward hinges. These evolutionary advances were the key to the great success of the dinosaurs. They did not have to throw the whole body from side to side to move their legs, so they could breathe easily while running quickly. They were able to grow bigger, walk further and move faster than any other reptiles. The two main groups of dinosaurs had different kinds of hips. The meat eaters and plant-eating sauropods (called saurischians, or lizard-hipped dinosaurs) had a forward-pointing pubis. In the plant-eating ornithischians, or bird-hipped dinosaurs, part of the pubis pointed backward, to allow more space for the gut.

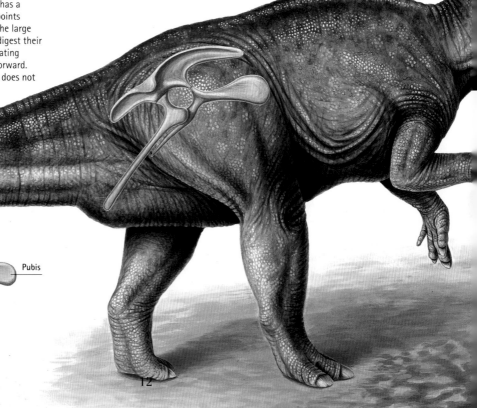

Ornithischian hip

Q: How did their hip bones affect the way dinosaurs developed?

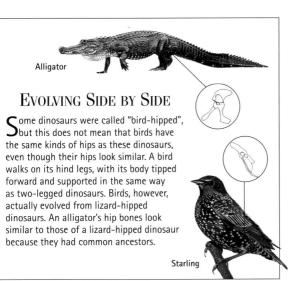

EVOLVING SIDE BY SIDE

Some dinosaurs were called "bird-hipped", but this does not mean that birds have the same kinds of hips as these dinosaurs, even though their hips look similar. A bird walks on its hind legs, with its body tipped forward and supported in the same way as two-legged dinosaurs. Birds, however, actually evolved from lizard-hipped dinosaurs. An alligator's hip bones look similar to those of a lizard-hipped dinosaur because they had common ancestors.

Alligator

Starling

STRANGE BUT TRUE

Fossils sometimes reveal illnesses, accidents and injuries. Sometime in the early Cretaceous Period, an *Iguanodon* fractured its hip. The dinosaur recovered, but was left with a bulge of new bone around the fracture.

Healed fracture

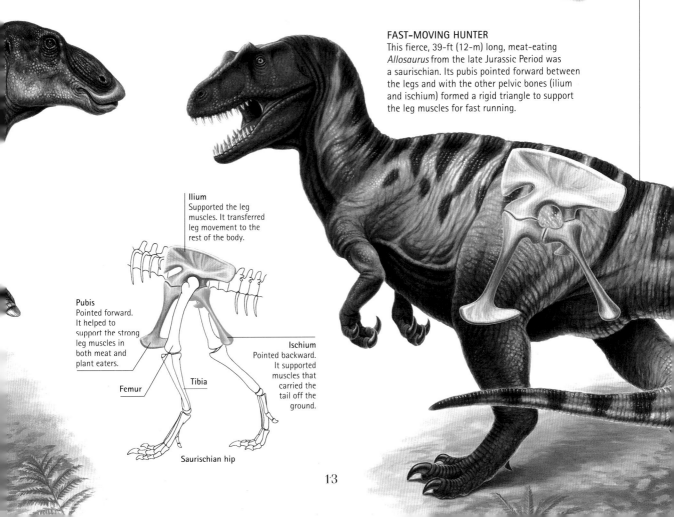

FAST-MOVING HUNTER
This fierce, 39-ft (12-m) long, meat-eating *Allosaurus* from the late Jurassic Period was a saurischian. Its pubis pointed forward between the legs and with the other pelvic bones (ilium and ischium) formed a rigid triangle to support the leg muscles for fast running.

Ilium
Supported the leg muscles. It transferred leg movement to the rest of the body.

Pubis
Pointed forward. It helped to support the strong leg muscles in both meat and plant eaters.

Ischium
Pointed backward. It supported muscles that carried the tail off the ground.

Femur

Tibia

Saurischian hip

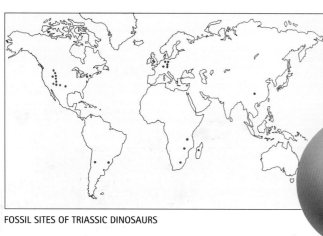

FOSSIL SITES OF TRIASSIC DINOSAURS

A VIEW OF THE WORLD
In the Triassic Period, the continents fitted together into one huge continent called Pangaea. Fossils show that most dinosaurs lived near the center of Pangaea, the area now divided among North America, Africa and northern Europe.

• THE ARRIVAL OF THE DINOSAURS •

The Triassic World

The Triassic Period was the "Dawn of the Dinosaurs." The Earth was a huge supercontinent called Pangaea (from the Greek meaning "all Earth"), which had three main environments and was dominated by mammal-like reptiles. Near the coasts, forests of giant horsetail ferns, tree ferns and ginkgo trees were alive with insects, amphibians, small reptiles (such as the first turtles, lizards and crocodiles), and early mammals. The dry, cool areas near the equator had forests of tall conifers (pine and fir trees) and palmlike cycads. The center of Pangaea was covered by hot, sandy deserts. Many different plants and animals developed in these varied climates. There was plenty of food for many life forms, especially one group of animals that first appeared 228 million years ago—the dinosaurs. These extraordinary creatures began to rule the Triassic world.

Ginkgo tree

Coelophysis

THE CHASE
In the warm, moist forest close to the coast of Pangaea, two *Coelophysis* chase a *Planocephalosaurus* up a tree.

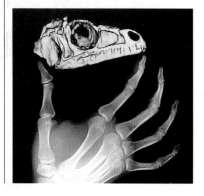

AN EARLY DINOSAUR
Eoraptor, the earliest known dinosaur (a fossilized head is shown supported by a human hand), lived 228 million years ago in what is now South America. This fast, lizard-hipped animal did not have a flexible jaw, so it could not trap struggling prey. It probably scavenged food from animals killed by larger reptiles.

STRANGE BUT TRUE
Coelophysis was an agile predator that used its strong, clawed hands to grab small prey. One fossil was found with the remains of a baby *Coelophysis* in its stomach. Did this Triassic hunter eat a member of its own species?

Q: What did Triassic dinosaurs eat?

TRIASSIC DINOSAURS

Zanclodon was a 20-ft (6-m) long, meat-eating carnosaur.

Herrerasaurus was a 10-ft (3-m) long, meat-eating coelurosaur.

Procompsognathus was a 4-ft (1.2-m) long, meat-eating coelurosaur.

Saltopus was a 2-ft (60-cm) long, meat-eating carnosaur.

Plateosaurus was an 26-ft (8-m) long, plant-eating prosauropod.

Horsetails

Cycads

DINOSAUR DIETS

Before flowering plants appeared in the Cretaceous Period, plant-eating dinosaurs grazed on ferns and tree leaves. Many plants had tough, waxy coatings or spines to protect themselves.

Meat-eating dinosaurs such as *Eoraptor* and *Coelophysis* hunted insects such as cockroaches and dragonflies, frogs, mammal-like reptiles and even early mammals —our distant ancestors.

Tree fern   *Wielandiella*   Dragonfly   *Haramiya*

Discover more in Meat-eating Dinosaurs

15

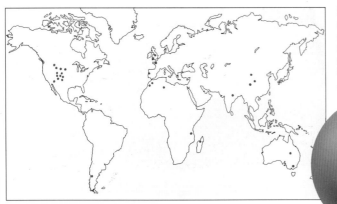

FOSSIL SITES OF JURASSIC DINOSAURS

MOVING APART
As the two supercontinents moved apart, the centers of dinosaur evolution spread. Most of the plant-eating sauropods, for example, remained in Gondwana, but some theropods, such as *Allosaurus*, spread throughout Laurasia.

> **STRANGE BUT TRUE**
> The forests that covered the Jurassic world survive today—as seams of coal. Dead trees (even whole trees destroyed by storms or floods) were covered by mud and soil. They slowly hardened into material that looks like rock, but burns like wood!

REARING ITS HEAD
A *Diplodocus* rears to defend itself against a predator. Its clawed front feet and lashing tail are ready for battle.

• THE ARRIVAL OF THE DINOSAURS •

The Jurassic World

Toward the end of the Triassic Period, the supercontinent of Pangaea started to divide into two smaller, but still very large, continents—Laurasia and Gondwana. New kinds of dinosaurs began to evolve on these new continents as they moved further apart. The slightly cooler temperatures and higher rainfall of the Jurassic world created a warm, wet climate that was ideal for reptiles. The lizard-hipped dinosaurs continued to break into the two groups that first split in the Triassic Period: the meat-eating theropods, which walked on two legs; and the plant-eating sauropods, which moved on all fours. The bird-hipped dinosaurs remained plant eaters. Giant, long-necked, plant-eating sauropods; plated, bird-hipped dinosaurs such as *Stegosaurus*; and bird-hipped plant eaters such as *Camptosaurus* were some of the mighty dinosaurs roaming the Jurassic world.

Camptosaurus, 20 ft (6 m) long, from Europe and North America

Allosaurus, 39 ft (12 m) long, from North America

Stegosaurus, 30 ft (9 m) long, from North America

Coelurus, 7 ft (2 m) long, from North America

Cycads

A SIZABLE APPETITE

Brachiosaurus was 39 ft (12 m) high and 75 ft (23 m) from nose to tail. It weighed an incredible 78 tons (80 tonnes)—as much as 12 African elephants—and ate the equivalent of 35 bales of hay a day. Its front legs were longer than its hind legs, and its whole body sloped downward from the shoulders (like a giraffe today), so its long neck could reach the tasty young leaves at the tops of the tallest trees.

FOOD FOR DINOSAURS
Jurassic dinosaurs ate many animals, including freshwater turtles such as *Pleisochelys* and perhaps even the earliest bird, *Archaeopteryx.*

Ground cover ferns

Discover more in Long-necked Dinosaurs

Q: Why were there so many plant-eating dinosaurs in the Jurassic Period?

FOSSIL SITES OF CRETACEOUS DINOSAURS

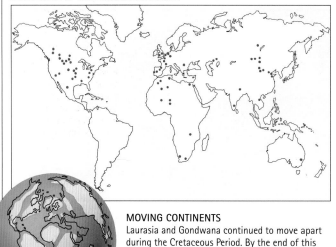

MOVING CONTINENTS
Laurasia and Gondwana continued to move apart during the Cretaceous Period. By the end of this period, the outlines of the continents were roughly the same as they are today. There were land bridges between the continents, but dinosaurs tended to evolve separately on each of the land masses.

• THE ARRIVAL OF THE DINOSAURS •

The Cretaceous World

The Cretaceous Period lasted for 80 million years. More dinosaur species evolved in this time than in all the other dinosaur periods put together. But at the end of this period, about 65 million years ago, the dinosaurs disappeared. The early Cretaceous Period was warm. Winters were mild and dry, and most of the rain fell in summer. Later, summers became hotter and winters were colder in the temperate and polar regions. The giant plant eaters disappeared and were replaced by smaller species such as *Triceratops* and the duckbilled dinosaurs. Flowering plants evolved during this period (giant plant eaters during the Jurassic Period ate and trampled down so much of the vegetation that it gave new plants the chance to grow) and were eaten by hundreds of new species of plant-eating animals. There was a huge amount of food to support an enormous number of animals. The animals that ate the flowering plants were also eaten by predators, from snakes (which first appeared in this period) to great predatory dinosaurs such as *Tyrannosaurus*.

Conifer forest

Magnolias

PREDATORS AND PREY
In this scene from the late Cretaceous Period in Mongolia, a *Velociraptor* (above right) battles with a *Protoceratops*, squashing dinosaur eggs in the process. An inquisitive *Prenocephale* looks on at the fierce encounter.

THE CYCLE OF LIFE
Flowering plants (**A**) were pollinated by insects (**B**), which were eaten by small mammals such as *Alphadon* (**C**), which in turn were eaten by dinosaurs such as *Dromaeosaurus* (**D**). Dinosaur droppings fertilized plants, and the cycle continued.

RELATED BUT NOT ALIKE
Animals that belong to the same family can evolve quite differently if they become isolated. *Hypacrosaurus* and *Bactrosaurus* were both duckbilled dinosaurs, and may have evolved from the same ancestor. But *Hypacrosaurus*, which lived in North America, was 30 ft (9 m) long and had a semi-circular crest on its head. *Bactrosaurus*, from central Asia, was only 13 ft (4 m) long. When the continents drifted apart, these dinosaurs evolved in different ways because they lived in such different places.

Q: Why did so many new dinosaur species develop during the Cretaceous Period?

Triceratops
Corythosaurus
Pachycephalosaurus
Saltasaurus
Euoplocephalus
Tyrannosaurus

• A PARADE OF DINOSAURS •

Meat-eating Dinosaurs

THE PREDATOR KING
Tyrannosaurus ("the tyrant lizard") was the largest predator ever to walk the Earth. This gigantic creature weighed more than an African elephant. Five complete specimens of *Tyrannosaurus* have been found around the world.

Many carnivorous dinosaurs were powerful, fast hunters that ate prey larger than themselves. But there were also smaller meat eaters (*Compsognathus* was no taller than a chicken) that ate eggs, insects, small reptiles and mammals. Carnosaurs, ceratosaurs and coelurosaurs, the three main groups of meat-eating dinosaurs, all had short, muscular bodies, slender arms, and low, powerful tails that balanced strong back legs with birdlike feet. With large eyes and daggerlike teeth, they were formidable predators: the coelurosaur *Deinonychus* had long, slashing claws on its front legs that could rip open a victim's belly; *Megalosaurus*, a carnosaur, had powerful hinged jaws that were armed with curved, saw-edged fangs. Meat-eating dinosaurs had much larger brains than plant-eating dinosaurs. Hunting prey that was large and sometimes armored required good vision and an ability to plan an attack.

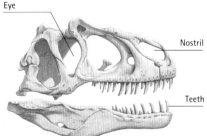

A MEAT-EATER'S SKULL
Allosaurus, a 39-ft (12-m) long predator from the late Jurassic/early Cretaceous Period, weighed more than a ton. Its skull could be up to 3 ft (1 m) long. As its jaws were hinged, *Allosaurus* could swallow large pieces of flesh whole.

FIGHTING FOR LIFE
Tenontosaurus, a plant eater from the early Cretaceous Period, fights a pack of ferocious *Deinonychus*. This reconstructed scene may well have happened. In the United States, the fossil skeleton of a *Tenontosaurus* was found surrounded by five scattered specimens of *Deinonychus*.

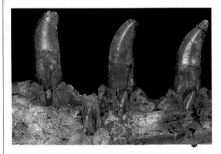

TEETH AND JAWS
Theropods, such as *Megalosaurus*, had long jaws that were usually armed with sharp, serrated teeth. New teeth were ready to replace those that wore out.

Compsognathus
Coelurosaur

Oviraptor
Coelurosaur

Albertosaurus
Carnosaur

Tools of Eating

The coelurosaur *Compsognathus* caught and ate prey with its hands. *Oviraptor* cracked open eggs with its beak. *Albertosaurus* had such short arms it had to tear off chunks of flesh with its powerful jaws. Scientists believe that the fish-eating *Baryonyx*, a theropod dinosaur discovered only in 1983, used the huge, hooklike claw on its hand to spear its prey.

Baryonyx
(not placed in a group)

Peg-shaped tooth | Leaf-shaped tooth

CUTTERS AND CHOPPERS
The shape of plant-eaters' teeth varied according to their diets. Dinosaurs that ate the hard leaves and fruit of cycads, palms and conifers had thick, peg-shaped teeth. Those that ate the leaves and fruit of softer, flowering plants had thinner, leaf-shaped teeth.

Triassic dinosaurs ate horsetail ferns as big as trees. | Jurassic dinosaurs ate pine cones and cycad fruits. | Cretaceous dinosaurs ate flowering plants such as magnolias.

STOMACH STONES
Sauropods, such as *Saltasaurus,* had no grinding teeth. They nipped off leaves with their slender, pencil-like teeth and ground them up with stomach stones called gastroliths. These were ground together by the muscular action of the stomach and crushed tough plant material.

DINOSAUR DINNERS
New varieties of plants evolved on the Earth along with new species of dinosaurs. The plant-eating sauropods, the biggest dinosaurs of all, had to eat huge quantities of plants to provide them with enough energy.

FINGER FOOD
Othnielia, a 5-ft (1.4-m) long gazelle-like dinosaur from the late Jurassic Period, used its five-fingered hands to push aside and hold down a fern while eating it. *Othnielia* had cheek pouches to store food so the tough plant material it ate could be chewed thoroughly later on.

• A PARADE OF DINOSAURS •

Plant-eating Dinosaurs

For most of the dinosaur age, the climate was warm and moist, and plants grew in abundance. Hundreds, perhaps thousands, of species of plant-eating dinosaurs grazed on the ferns, cycads and conifers of the Triassic and Jurassic periods, then on the flowering plants of the Cretaceous Period. The bird-hipped ornithopods (the pachycephalosaurs, iguanodonts, duckbills, armored dinosaurs, and the horned dinosaurs that roamed in huge herds during the late Cretaceous Period) had special cheeks to store plants while they were busy chewing. The lizard-hipped sauropods, which included *Apatosaurus, Diplodocus* and *Brachiosaurus,* reached the lush vegetation at the tops of the trees with their long necks. These enormous plant eaters had large fermenting guts and stomach stones (as discussed above) that helped them digest huge amounts of plants.

Teeth and Beaks

Paleontologists can tell much about how a dinosaur lived from the shape of its teeth or its beak, if it had no front teeth. Giraffes' teeth are different from zebras' teeth because giraffes eat tender leaves from the tops of trees, while zebras eat tough, dry grass. In the same way, different families of dinosaurs evolved different kinds of teeth and beaks to cope with a variety of plants.

Protoceratops, one of the smaller horned dinosaurs, had a parrotlike beak for shearing off plant stems, and scissorlike teeth to slice up its food.

Camarasaurus, an 59-ft (18-m) long sauropod, had spoon-shaped cutting teeth but no grinding teeth. However, it could reach high into trees to tear away leaves.

Corythosaurus, a duckbill, tore off leaves with its horny beak, stored them in its cheek pouches, then used rows of strong interlocking teeth to grind them.

Plateosaurus, an early, long-necked giant, had leaf-shaped teeth to pluck off the leaves of soft plants such as ferns. It did not have grinding teeth.

Iguanodon grazed on tough plants such as horsetails, and used its horny beak to nip off leaves. Its rows of ridged, grinding teeth crushed the leaves into a pulp.

Strange but True

Heterodontosaurus, a plant eater from the early Jurassic Period, had three kinds of teeth. In the front upper jaw it had small cutting teeth; on the lower jaw it had a horny beak. Then it had two pairs of large, fanglike teeth, with grinding teeth at the back.

• A PARADE OF DINOSAURS •

Long-necked Dinosaurs

LONGEST NECK
Mamenchisaurus had the longest neck of any known animal—an amazing 36 feet (11 m). It could hardly bend its neck, but it could rear up on its hind legs to reach the tops of the highest trees.

The biggest, heaviest and longest land animals that have ever lived were the long-necked sauropod dinosaurs. These strange creatures had long tails, compact bodies, small heads, front legs that were shorter than the hind legs, and clawed first fingers or thumbs that were much larger than their other fingers (they may have used these to hook branches). In 1986, paleontologists unearthed a few bones from an enormous sauropod named *Seismosaurus* ("earthquake lizard") that may have been more than 98 ft (30 m) long. They have also discovered complete skeletons from sauropods almost as large. *Brachiosaurus*, for example, grew to 75 ft (23 m), stood 39 ft (12 m) high and weighed as much as 12 African elephants. *Brachiosaurus* and the other giants had pillarlike legs to support their great weight, but their skeletons were very light. Their bodies were shaped like giant barrels, and they carried their long tails high off the ground. *Diplodocus'* tail ended in a thin "whip;" other sauropods may have had tail clubs for self-defense.

MIGRATING HERDS
Trackways (fossil footprints) and groups of fossils indicate that many sauropods lived in herds and may have migrated to find fresh food, with the adults protecting their young from predators.

Small head
A small head and a small mouth meant *Diplodocus* had to spend a lot of time eating to nourish its huge body.

Strong but light
Struts of bone and air spaces kept *Diplodocus'* skeleton light but strong.

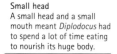

Long legs
With the help of its long front legs, this grazer could reach the tender young leaves in the treetops.

LONGER BUT LIGHTER
Diplodocus was longer than *Brachiosaurus*, but weighed only a third as much—partly because the skeleton of *Diplodocus* contained air spaces that reduced its weight but not its strength. The head of a *Diplodocus* was no larger than that of a horse, but its neck was 23 ft (7 m) long and its tail stretched an incredible 46 feet (14 m).

Strong legs
Diplodocus' hind legs were almost solid pillars of bone to support the weight of its intestines and tail muscles.

LONGER NECK
Brachiosaurus carried its tiny head high off the ground on the end of a 20-ft (6-m) long neck. Its front legs were almost as long as its hind legs, which gave its already long neck extra reach.

Strange but True
Although sauropods did not need water to support their weight, apparently they could swim. One trackway seems to show *Diplodocus* walking only on its front feet. It must have been floating in water, pushing itself along with its front feet.

Fit for a King
In 1905, American millionaire Andrew Carnegie presented a plaster cast of *Diplodocus* to the Natural History Museum in London. He had financed the excavation of the original specimen. King Edward VII was there to unwrap the biggest present any king has ever received! Ten copies of the skeleton were sent to other museums around the world.

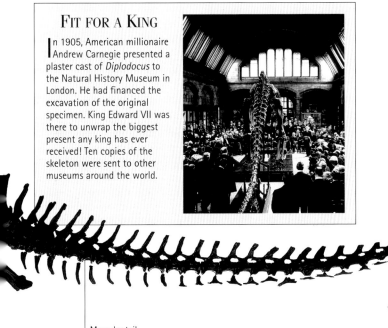

Whiplash
Diplodocus could injure or stun predators with the bony tip of its 46-ft (14-m) long tail.

Muscular tail
Diplodocus could not outrun a predator, but its great size and heavy, strong tail protected it.

LONG NECK
A modern giraffe, like all mammals, has only seven neck vertebrae. Sauropods, however, had between 12 and 19 neck vertebrae—all with bony struts to provide extra support.

• A PARADE OF DINOSAURS •

Armored, Plated and Horned Dinosaurs

Many plant-eating dinosaurs evolved in strange ways to defend themselves against predators or to fight over mates and territories. Pachycephalosaurs, for example, had domed-shaped heads with thick, strong layers of bone. The plated and armored stegosaurs were slow-moving, small-brained ornithischians that relied on spikes and armor plating to defend themselves. The best known are *Stegosaurus*, a 30-ft (9-m) long, late Jurassic dinosaur that had one or two rows of plates along its back, and two to six pairs of long, sharp spikes at the end of its strong tail; and *Ankylosaurus*, a 33-ft (10-m) long, late Cretaceous dinosaur that was protected by hundreds of bony nodules (some of them with spiky bumps) on its back and sides and a double-headed club of bone on its tail. The ceratopians, or horned dinosaurs, were the last group of ornithischians to evolve before the dinosaurs died out at the end of the Cretaceous Period. They lived for only 20 million years or so, but spread out across North America and Asia. Ceratopians formed vast herds and used their platelike, horned heads to protect themselves and their young against predators such as *Tyrannosaurus* and *Velociraptor*.

PREDATORS BEWARE!
Styracosaurus, a 16-ft (5-m) long ceratopian, defends its young from a predator by displaying its nose horn and spiked head shield. Its spiky frill protected its neck, and it could use its nose horn to rip open a predator's belly.

BIG BUT LIGHT
Chasmosaurus' head shield was light and easy to move. It was designed more for display than defense, since this animal could easily outrun predators.

SMALL BUT STRONG
Centrosaurus, a 20-ft (6-m) long ceratopian, was a slow-moving animal that defended itself with its short, heavy head shield.

26

BODY ARMOR
Kentrosaurus, a 16-ft (5-m) long stegosaur from Africa, had seven pairs of plates from the neck to the middle of the back, and seven pairs of spikes on its back, hips and tail.

SPIKES AND SPINES
Polacanthus, a 13-ft (4-m) long nodosaur, protected its head and vital organs with a double row of vertical spines and used its strong, spiked tail for self-defense.

DISHES AND DAGGERS
Stegosaurus may have used its back plates for defense, or for heating and cooling, but its tail spikes were used for self-defense. They were fused to the bones of the tail, so *Stegosaurus* could swing its spiky tail to scare off a predator.

BUILT FOR DEFENSE
An ankylosaur *Euoplocephalus* from the late Cretaceous Period moved slowly and had a small brain. It could not hope to outwit fast, intelligent predators such as *Velociraptor*. But 20-ft (6-m) long *Euoplocephalus* was very heavily armored—even its eyelids were protected by bony shutters. It could cause serious damage with the club at the end of its 8-ft (2.5-m) long tail.

Q: Why would it be unwise for *Tyrannosaurus* to attack a horned dinosaur?

A HOLLOW CREST
A male *Parasaurolophus* could stay in touch with other members of the herd or bellow a challenge to another male by forcing air from its mouth up into its hollow crest, then out through its nostrils. It must have had flaps or valves inside the crest to stop it from hooting whenever it breathed.

• A PARADE OF DINOSAURS •

Duckbilled Dinosaurs

The duckbilled dinosaurs (hadrosaurs) had broad, ducklike beaks. They walked or ran on their hind legs, and leaned down on their shorter front legs to graze on vegetation. There were many species of duckbills; they were the most common and widespread plant-eating dinosaurs of the late Cretaceous Period. Hadrosaurs probably evolved in central Asia, but spread to Europe and North America. They had a varied diet, which meant they were able to survive as the Cretaceous climate became drier. All hadrosaurs were closely related, but they looked very different from each other. Some may have had inflatable nose sacs so they could communicate with each other by hooting. Others had hollow crests that acted like echo chambers. They could bellow or call each other by making noises that may have sounded like those made by a modern bassoon (above left).

Q: Why did duckbill dinosaurs hoot?

HEAD OF THE FAMILY

Paleontologists used to believe that duckbills with different crests belonged to different species. Now they think these different-crested duckbills were members of the same species. A female *Parasaurolophus*, for example, had a medium-sized, curved crest; a young, or juvenile *Parasaurolophus* had a short, fairly straight crest; and an adult male had a long, curved crest. All used their hollow crests to produce sounds that other members of the herd would understand.

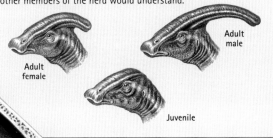

Adult female
Adult male
Juvenile

A DUCK'S BILL
Like all hadrosaurs, 43-ft (13-m) long *Edmontosaurus* had a toothless duckbill, covered with leathery skin, which it used to pluck leaves and fruits. It had rows of teeth in the back of its mouth, and it chewed food by moving its jaw up and down so the overlapping teeth crushed its food.

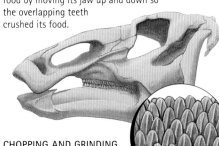

CHOPPING AND GRINDING
Seen close up, *Edmontosaurus'* tooth rows consisted of scores of tiny, leaf-shaped teeth, which acted like a cheese grater.

DID YOU KNOW?
Saurolophus, which had only a small, hornlike crest, may have produced noises by inflating a skin-covered sac on top of its nose. This would have been supported by the crest at the back of its head.

LIVING TOGETHER
Like giraffes (which eat tree leaves) and zebras (which eat low-growing plants), flat-headed and crested duckbills were able to live together without taking one another's food supply.

Discover more in The Cretaceous World

• A PARADE OF DINOSAURS •

Record-breaking Dinosaurs

The dinosaurs were one of the world's most successful group of animals. They were the biggest, heaviest and longest land animals that have ever lived, and they dominated the Earth for almost 150 million years. Compared to this record, the four million years that humans have been on Earth seems like the blink of an eye. Dinosaurs were the world's strangest and most extraordinary animals. It has always been hard for people to imagine a world populated by such huge creatures: the remains of the first dinosaur ever described, 30-ft (9-m) long *Megalosaurus*, were first thought to belong to a human giant. It has also been hard for people to understand just how spectacular the dinosaurs were. *Seismosaurus*, the "earthquake lizard" and the biggest of the sauropods, may have been more than 98 ft (30 m) long. Only a few of its bones have ever been found: a 8-ft (2.4-m) long shoulder blade, taller than the biggest human giant; and a 5-ft (1.5-m) long vertebra. The ground must have quaked with each footstep from this gigantic creature.

SMALLEST
Compsognathus was one of the smallest known dinosaurs. It was only 3 ft (1 m) long, weighed just 8 lb (3.5 kg), and stood no taller than a chicken. It must have been a swift and efficient hunter. One specimen was found with the bones of a tiny lizard in its stomach cavity.

HEAVIEST
Weighing 78 tons (80 tonnes), 75-ft (23-m) long *Brachiosaurus* was as tall as a four-storey building. Its shoulders were more than 20 ft (6 m) off the ground and its humerus, or upper arm bone, was 7 ft (2 m) long. The humerus of an adult human is only about 1 ft (35 cm) long.

DID YOU KNOW?
What do a *Struthiomimus* (whose name means "ostrich-mimic") and an ostrich have in common? They can both sprint swiftly on very long slim legs (an ostrich can outrun a horse) and have long thin necks with small heads.

FASTEST
Struthiomimus, stood 7 ft (2 m) high and was 10–13 ft (3–4 m) long. It defended itself against predators by running at speeds of up to 31 miles (50 km) per hour, balancing on its long, birdlike hind legs.

LONGEST NECK
Mamenchisaurus, at 72 ft (22 m), was almost as long as its close relative *Diplodocus*, but it had a fairly short tail. Its 36-ft (11-m) long neck, which it used to reach the tops of tall trees, is the longest neck of any known animal.

LONGEST
With more than half of its total length of 89 ft (27 m) taken up by its 46-ft (14-m) long tail, *Diplodocus* was the longest known dinosaur. It would have used its strong, whiplike tail to defend itself against predators such as *Allosaurus*.

BIGGEST PREDATOR
Tyrannosaurus was bigger than any predator except the sperm whale. It could grow up to 46 ft (14 m) long and was taller than a double-decker bus. It weighed 7 tons (7 tonnes).

Big, Bigger, Biggest

In the 1970s and 1980s, fossil hunters found massive bones from sauropods even bigger than 75-ft (23-m) long *Brachiosaurus*. Called *Supersaurus, Ultrasaurus,* and *Seismosaurus*, these incredible animals may have been 98 ft (30 m) long! In this photograph, paleontologist Dr. James Jensen stands next to the reconstructed front leg of one of these giants. These fossils are still being unearthed, and it may take 10–20 years to reconstruct their skeletons. Then they will topple the record-breakers of today.

MOST TEETH
Anatotitan, a duckbilled dinosaur, had about 1,000 tiny, leaf-shaped teeth arranged in rows of 200–250 on each side of its upper and lower jaws, all at the back of its mouth. Two mummified fossils of this species have been found, complete with the remains of their last meals: pine needles, twigs, seeds and fruits.

Q: Will these dinosaurs always be record-breakers?

• UNCOVERING DINOSAUR CLUES •

Fossilized Clues

We rely on fossils for clues about how dinosaurs lived. But dinosaur fossils are very rare. The chances of a plant or animal becoming fossilized were low because conditions had to be just right for fossilization to occur. An animal had to be fairly big (small dinosaurs had delicate bones that were easily scattered or destroyed, or eaten by scavengers), and it had to die in the right place. If a dinosaur's body was washed into a lake, for example, silt would cover it up quite quickly and its bones were more likely to be preserved. In most cases only the bones of the dinosaurs were preserved (a few turned into minerals such as opal), but occasionally the animal was covered by sand or volcanic ash that preserved, or mummified, the body and left an impression of the texture of its skin. Sometimes, only a dinosaur's footprints or its droppings have been preserved. Paleontologists use all these clues to piece together pictures of the creatures that lived so many millions of years ago.

BACK IN TIME
The deeper a layer of rock, the older it is. The oldest and deepest rocks contain single-celled bacteria and algae. More complex plants and animals are found in the newer rocks above.

A FOSSIL IN THE MAKING
A 20-ft (6-m) long *Camptosaurus* lies at the water's edge, dead of disease or old age. The hot sun has begun to dry the body, and if scavengers do not tear it apart, it will be covered by silt and gradually fossilized. The *Coelurus* shown here are eating the insects and other animals around the carcass. Their jaws are too weak for the thick skin of *Camptosaurus*.

Q: Why are there so few dinosaur fossils?

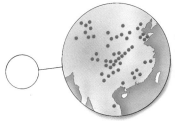

MONGOLIAN FOSSIL SITES
Mongolia, in central Asia, is covered by sand and desert today. During the Jurassic and early Cretaceous periods, Mongolia was warm and moist, with lakes and shallow seas. Many species of dinosaur lived in this ideal environment.

NORTH AMERICAN FOSSIL SITES
North America was warm and moist in the Jurassic and early Cretaceous periods. Great herds of plant eaters lived there, but they disappeared when the climate became colder and more changeable toward the end of the Cretaceous Period.

BONES THAT ARE NOT BONES

Skeletons can fossilize in different ways. In a petrified fossil (right), preserved bones that form partial or complete skeletons have outer and middle layers that have been replaced by minerals. They literally "turn to stone." A cast is formed when mud fills the hollow spaces inside bones—this occasionally happens with a dinosaur's brain or the canals of its middle ear. The bone then rots away. Sometimes fossil skulls are hollow, and scientists can make a mold of the dinosaur's brain. Very rarely, a dinosaur is mummified in dry sand that gradually hardens into rock, leaving an impression of the animal's skin.

OUT OF REACH
Beneath the surface of a lake, a dead dinosaur is safe from large scavengers. Its flesh rots away or is eaten by fish, and the skeleton remains intact.

COVER-UP
Layers of sand or silt cover the dinosaur's bones, and stop them from being washed away.

FOSSILIZATION
Trapped and flattened by layers of sediment, the dinosaur's bones are gradually replaced by minerals that are harder than the rocks around them.

FOSSIL FINDS
Millions of years later, upheavals in the Earth's crust bring the dinosaur's fossilized skeleton close to the surface, where it is exposed by the weather and erosion.

Discover more in Skeletons and Skulls

• UNCOVERING DINOSAUR CLUES •

Skeletons and Skulls

Most of the dinosaurs we know about were much bigger than even large modern mammals. An average-sized dinosaur such as *Camptosaurus* was about 21 ft (6.5 m) long—a third longer than an African elephant. But *Camptosaurus* weighed only about 3 tons (3 tonnes)—less than half the weight of an elephant. Dinosaurs had two distinct body types: a bipedal fast-running kind such as *Hypsilophodon*, and a quadrupedal heavy type such as *Camarasaurus*. They could grow to enormous sizes because their skeletons were superbly engineered; they were very strong without being very heavy. The vertebrae of the giant sauropods were supported by struts and thin sheets of bone because they were almost hollow; solid vertebrae would have made these animals too heavy to stand upright. Most dinosaurs had holes in their skulls. Meat eaters had the largest holes of all, to accommodate the bulging and powerful jaw muscles that opened and closed their jaws.

Skull
This 5-ft (1.5-m) long, lightly built "gazelle" of the dinosaur world had a horny beak at the front of its mouth, a fairly large brain and large openings for the eyes.

Backbone
Extra ribs in front of the shoulders supported the neck muscles.

HYPSILOPHODON

Hands
Four long, clawed fingers were used to grasp plant food or to support *Hypsilophodon* as it grazed on low-growing plants.

CAMARASAURUS

Backbone
Like the steel rods of a crane, the vertebrae provided support where it was needed most.

Skull
This 59-ft (18-m) long sauropod had a small head. The large openings on top of the skull may have helped to cool the small brain.

Leg bones
Like all sauropods, *Camarasaurus* had massive, pillarlike legs to carry its great weight.

Chest
Very deep ribs supported the large stomach *Camarasaurus* needed to digest its tough plant food.

Hind feet
Camarasaurus's toes could spread out for support as it reared up to reach young leaves in the treetops.

Front feet
Five strong, clawed toes helped to support the weight of the chest, neck and head.

CERATOSAURUS
This 20-ft (6-m) long predator had a strong lower jaw and a high, narrow skull. Both had large cavities to make space for the huge jaw muscles that drove razor-sharp fangs.

OURANOSAURUS
Although its jaw muscles were weak, *Ouranosaurus* was an efficient plant feeder. As it closed its mouth, the bones of the upper jaw moved apart, breaking up food with bands of cheek teeth.

DINOSAUR BRAINS

Dinosaurs had small brains, but this does not mean they were stupid. Brain size is not the most important factor. Brain complexity and brain size in relation to total body size are far more important. *Iguanodon's* body was as big as a bus; its brain was no bigger than a goose's egg. *Iguanodon* did not need much intelligence to find its food of leaves and fruit. *Deinonychus*, however, had to see, smell and chase fast-running animals. Its brain was the size of an apple, even though its body was smaller than a car.
But the "thinking" part of a dinosaur's brain, the cerebrum, was much smaller than a mammal's cerebrum, which meant that a dinosaur could not have learned new things as easily as a monkey or a dog could today.

Rhesus monkey

Iguanodon

Q: What determined the size of dinosaurs' brains?

Tail
Spines on the undersides of the tail vertebrae supported muscles that helped *Hypsilophodon* carry its tail off the ground.

Leg bones
Hypsilophodon was a fast runner. As its femur was very short, the long tibia and foot could swing forward to give the dinosaur great speed.

Feet
Long, strong toes like those of an ostrich gave *Hypsilophodon* a sure footing as it sprinted away from danger.

IN THE NOSE
Corythosaurus (left) may have drawn air in through the nostrils at the front of its snout and into its hollow crest to produce sounds. *Brachiosaurus* had nostrils on the top of its head, possibly to help keep its body cool.

DID YOU KNOW?
Stegosaurus's brain was the size of a walnut. A cavity in the vertebrae above its hips may have housed a gland that produced energy-rich glycogen. This gave *Stegosaurus* a burst of speed if it needed to escape from a predator.

BUILT FOR STRENGTH
The tibia and femur of *Tyrannosaurus* were the same length, and had powerful muscles attached to them. *Tyrannosaurus* could charge at its prey with a sudden burst of energy, but its legs were not designed for a long chase.

Discover more in Plant-eating Dinosaurs

• UNCOVERING DINOSAUR CLUES •

Footprints and Other Clues

ARMOR PLATING
Dinosaur skin, like that of living reptiles, was made up of scales, sometimes with bony lumps (called osteoderms) that provided protection against predators' teeth.

TRACKING FOSSIL FOOTPRINTS
Footprints show that many dinosaurs traveled in groups. These *Apatosaurus* prints were made by five adults moving in the same direction.

DAILY DIET
Coprolites (fossilized dung) have been found containing hard seeds, pieces of pine cones, and even plant stems.

Fossilized teeth and bones tell us much about how dinosaurs looked and lived. But paleontologists also use other clues to piece together pictures of the dinosaurs' day-to-day lives. Skin impressions show that dinosaurs were protected against predators and spiky plants by a tough covering of skin. Fossil footprints, called trackways, tell us how dinosaurs moved about, and that sauropods, hadrosaurs and horned dinosaurs traveled in herds. The remains of nests show that dinosaurs built nests close to each other for protection against predators and scavengers. The fossils of eggs and even baby dinosaurs indicate how small these animals were when they hatched and how quickly they grew. The bones of adult dinosaurs give clues to their diet, injuries and the cause of their death, while fossilized dung provides information about what dinosaurs ate.

STORIES IN STONE
Small plant eaters stick close to a herd of long-necked sauropods as it migrates across the late Jurassic landscape of North America. Sharp-clawed theropods shadow the herd, hoping to pick off a sick or injured animal.

True Colors

We will never know what color the dinosaurs were. Fossilized skin does not preserve colors, so the colors and markings we give dinosaurs are those from our imaginations. Some paleontologists believe plant-eating dinosaurs had dull, dark colours so predators could not see them. Meat-eating dinosaurs may have been dull colored too, so they could hide and ambush plant eaters. Other scientists think plant-eating dinosaurs (especially males) changed color at different times of the year (as imagined in these illustrations of the duckbill *Lambeosaurus*) to mate or to defend their territories.

FOOTPRINT CLUES
Scientists can estimate how quickly each dinosaur was moving by calculating the length of its pace, the length of its stride and the length of the animal's feet and legs.

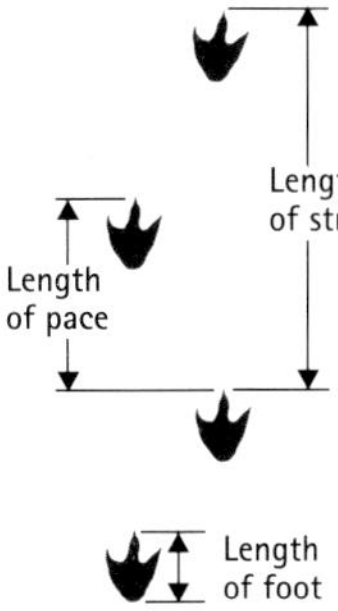

Length of stride
Length of pace
Length of foot

Q: What color would you choose for *Tyrannosaurus*?

ON GUARD
The *Maiasaura* nests discovered by Dr. Horner were spaced about 23 ft (7 m) apart, the same length as an average adult, so *Maiasaura* had enough space to avoid accidentally crushing one another's eggs. However, they were close enough to protect the nests from possible egg thieves such as *Oviraptor*.

ON THE INSIDE

Amniotic sac
A fluid-filled bag cushioned the embryo.

Chorion
This membrane provided oxygen.

Yolk sac
This provided nourishment.

Eggshell
Dinosaur young developed inside a sealed container.

A DINOSAUR NURSERY
Up to 25 eggs were laid in each *Maiasaura* nest, which was a 7-ft (2-m) wide, 3-ft (1-m) deep bowl scooped out of mud. The hatchlings were about 1½ ft (50 cm) long.

• LIFE AS A DINOSAUR •

Raising a Family

Paleontologists used to think that dinosaurs did not look after their eggs or their young because very few dinosaur nests had been discovered. In 1978, however, Dr. John Horner found a duckbill dinosaur nesting site in North America, with dozens of nests spaced just far enough apart so that adult dinosaurs could guard their own eggs without stepping on another dinosaur's nest. He also found fossil eggshells and the fossils of 15 baby duckbills. The babies had already grown much larger than when they were born, but they had not left the nest because they were still being cared for by their parents. Dr. Horner called these dinosaurs *Maiasaura*, or "good mother lizards."
    Paleontologists know that at least two meat eaters, *Troodon* and *Oviraptor*, laid eggs. Fossil eggs from the giant sauropods have been found in Europe, South America and China, but we do not know how these enormous creatures managed to lay their eggs safely.

Chicken's egg

Possible theropod egg

Oviraptor's egg

Emu's egg

Q: Why was Dr. Horner's discovery so important?

BIG BODIES, SMALL EGGS

Dinosaur eggs were very small in proportion to their bodies. Very large eggs would have very thick shells, and these could never be broken by the hatchlings.

A TERRIBLE EGG THIEF?

Many paleontologists believe that *Oviraptor*, a theropod from the late Cretaceous Period, stole eggs from the nests of other dinosaurs. Its strong jaws could have easily broken eggshells and crushed the bones of the young dinosaurs it caught with its clawed hands. The first *Oviraptor* fossil, discovered in Mongolia in 1924 with a clutch of eggs, seemed to confirm this belief. Paleontologists thought the eggs belonged to a *Protoceratops*, but new evidence has shown that the eggs did in fact belong to *Oviraptor*. The debate continues.

DID YOU KNOW?

Microscopic examination of *Maiasaura* embryos and hatchlings shows that they had very poorly developed joints in their legs. They had to be cared for by their parents. Hypsilophodons (the cousins of hadrosaurs such as *Maiasaura*), however, had strong legs and could fend for themselves as soon as they hatched.

Discover more in Duckbilled Dinosaurs

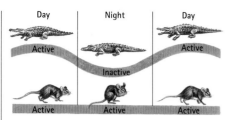

UPS AND DOWNS
A cold-blooded reptile, such as a crocodile, must bask in the sunlight before it is warm enough to move quickly. But a warm-blooded mammal, such as a mouse, can be active all the time. Its body temperature stays the same no matter how cold it is outside.

Staying cool
Spongy skin over the sail would have allowed heat to radiate quickly. Even a small drop in the temperature of the blood would have helped *Ouranosaurus* to stay cool.

Cooling the blood
A complex system of small veins carried warm blood up into *Ouranosaurus'* sail, where it was cooled before it flowed back down into the body.

• LIFE AS A DINOSAUR •

Keeping Warm; Keeping Cool

Every animal needs to keep its body temperature stable. If it cannot stay warm, it will not have enough energy to move around. If it becomes too warm, its brain may overheat and its breathing and digestion will not work properly. Dinosaurs had different ways to keep their bodies at stable temperatures. Some had frills, plates, sails or spikes to help them warm or cool their blood. Some scientists believe that dinosaurs may have had divided hearts to pump blood into their brains. This advanced body structure would probably have included an internal temperature control, or warm-bloodedness. The small, energetic meat eaters may have been warm-blooded, but this does not seem to be true for large, adult dinosaurs. Their bodies had a wide surface area, which meant they were able to radiate heat quickly and maintain a stable temperature.

AIR CONDITIONING
Ouranosaurus could warm up quickly in the morning, then cool down by the afternoon by pumping blood under the skin of the spiny "sail" on its back.

Q: How did dinosaurs keep cool?

HOT PLATES
Tuojiangosaurus had 15 pairs of bony plates along its back. Some paleontologists think the plates were covered by skin that was rich with blood to help the dinosaur warm up or cool down quickly.

BIVALVES
A divided heart separates blood that flows under high pressure to the body from blood that flows under low pressure to and from the lungs. Some scientists think that tall or fast-moving two-legged dinosaurs needed a high-pressure blood supply to their brains and muscles.

Low pressure   High pressure

NECKS AND TAILS
Sauropods had long necks to help them reach the treetops, and long tails to balance their long necks. Their necks and tails provided a large surface area to soak up heat from the sun or to cool them down in the middle of the day.

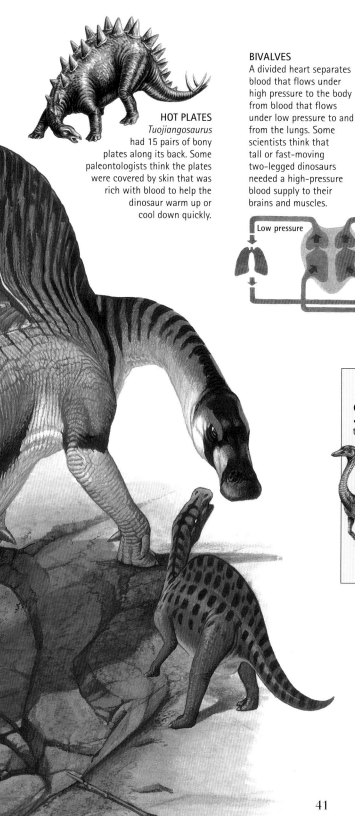

A Change in Temperature

Scientists have looked at dinosaur bones under a microscope, and believe these reptiles may have been both warm-blooded and cold-blooded. Dinosaurs seem to have grown quickly (like warm-blooded mammals) when they were young, then more slowly (like cold-blooded reptiles) when they became adults. But small meat eaters, such as *Dromiceiomimus* (left), were very active predators, hunting lizards and insects. They would have needed a constantly stable temperature and may have been more warm-blooded as adults than other dinosaurs.

Nostril

Strange but True
Sauropods such as *Brachiosaurus* could "let off steam" when they became too hot by pumping blood through the delicate skin inside their huge nostrils. This cooled the blood so the rest of the body could then keep cool.

Discover more in What is a Dinosaur?

A WIDE VARIETY
Like a duckbilled dinosaur, this rhinoceros has a wide mouth so it can eat many types of plants.

SPECIAL DIETS
Like a horned dinosaur, this gazelle chooses its food carefully, plucking leaves and fruit with its narrow mouth.

Into the mouth
Paleontologists do not know if *Apatosaurus* had a muscular tongue or used its peglike teeth to rake leaves or twigs into its mouth.

Down the neck
Powerful muscles pushed food down *Apatosaurus*' 20-ft (6-m) long esophagus, a tube running from the mouth to the stomach.

Living reptile
Crocodile teeth are designed to grip, not cut.

• LIFE AS A DINOSAUR •

Eating and Digesting

Different dinosaurs ate and digested their food in various ways. Scientists have learned about these dietary habits by studying dinosaur teeth and bones, analyzing dinosaur dung and observing how living animals eat and digest their food. Paleontologists have found fossils of dinosaurs that sliced and ripped meat, dinosaurs that chewed plants or ground leaves into a paste before swallowing them, and toothless dinosaurs that ate eggs. The meat eaters had sharp teeth to cut up meat, which is easier to digest than coarse plants. *Tyrannosaurus*' sharp, serrated teeth were designed so that its prey's struggles actually helped it tear off chunks of flesh. Large plant-eating dinosaurs had internal features such as stomach stones (gastroliths) to help grind and digest the large quantity of plants they ate.

Large theropod
Tyrannosaurus tooth

Small theropod
Troodon tooth

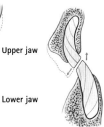

Upper jaw

Lower jaw

CUTTING DOWN TO SIZE
Styracosaurus used rows of scissorlike teeth to snip leaves into small pieces.

GRINDING TO A PASTE
Edmontosaurus used rows of grinding teeth to crush leaves into a paste.

DID YOU KNOW?

A fossil of a large plant-eating sauropod has been found with 64 large, polished stones inside its ribs, which is where the stomach would have been when the dinosaur was alive.

DINOSAUR DUNG

The dung of plant-eating dinosaurs was often hard enough to become fossilized. Paleontologists study fossil dung, or coprolites, to figure out what kinds of plants different dinosaurs ate. Coprolites can tell us how the sauropods grew to be so large before flowering plants evolved, and how large herds of duckbills could survive in the fairly dry conditions of the Cretaceous Period.

Q: How do paleontologists know what dinosaurs ate?

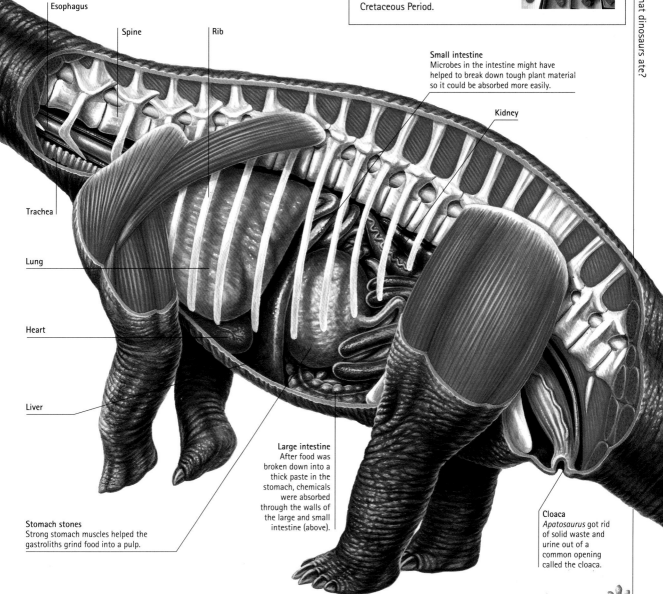

Esophagus

Spine

Rib

Small intestine
Microbes in the intestine might have helped to break down tough plant material so it could be absorbed more easily.

Kidney

Trachea

Lung

Heart

Liver

Stomach stones
Strong stomach muscles helped the gastroliths grind food into a pulp.

Large intestine
After food was broken down into a thick paste in the stomach, chemicals were absorbed through the walls of the large and small intestine (above).

Cloaca
Apatosaurus got rid of solid waste and urine out of a common opening called the cloaca.

Discover more in Footprints and Other Clues

A SPIKY SHIELD

Triceratops' neck was a massive frill of solid bone with 3-ft (1-m) long horns that protected its neck and chest from an attack by another *Triceratops* or a predator.

BUILT LIKE A TANK

Euoplocephalus was protected by bands of armor, bony studs on the shoulders and a heavy, bony skull. It could injure a predator by lashing out with a bony club at the end of its tail.

STABBING TAIL

To defend itself against a predator, *Tuojiangosaurus* used its muscular tail, which was armed at the tip with two pairs of sharp spikes.

DID YOU KNOW?

Iguanodon used its hand in many ways: for walking, for grasping food, for stripping leaves from branches and for self-defense. This peaceful plant eater could also use its thumb spike to injure or kill a predator by stabbing its neck or eyes.

MULTI-PURPOSE TAIL

Diplodocus' tail was very long. It used the tail for support when it reared up to crush a predator with its front legs or swung it like a whip to blind or stun an attacker.

• LIFE AS A DINOSAUR •

Attack and Defense

Many dinosaurs used their horns, spikes or armor to defend themselves. But even those without armor had their own defense weapons. *Apatosaurus* could rear up on its hind legs and crush an attacker with its front feet, or use its tail to injure a predator. Many sauropods travelled in herds, relying on safety in numbers so that only weak or sick animals would be attacked. The bird-mimic dinosaurs such as *Gallimimus* used their speed to escape, while *Pachycephalosaurus* could use its thick skull to defend itself against predators and other members of its own species. Meat eaters had speed, agility and sharp teeth for effective attack and defense. Large predators such as *Tyrannosaurus* hunted alone, and relied on a surprise rush. We will never know if dinosaurs used camouflage. Perhaps some species of plant eaters had dappled skin so they could hide from predators. Meat eaters may have used the same kind of disguise to ambush their prey.

THE TERRIBLE CLAW

Just as a falcon uses its razor-sharp claws to kill its prey, *Deinonychus* (whose name means "terrible claw") used the 5-in (13-cm), swivel claw on the second toe of each foot to kill its prey. It would leap into the air to kick or balance on one leg as it slashed at the skin of plant eaters. Fossils of five *Deinonychus* have been found beside the body of a *Tenontosaurus*, which suggests that the fast-moving, big-brained *Deinonychus* hunted in packs.

BATTERING RAM
Pachycephalosaurus' skeleton was designed to withstand its charging attacks against other males, or predators.

HEAD TO HEAD
Two 26-ft (8-m) long male *Pachycephalosaurus* are butting heads like mountain goats to see which will mate with a herd of females. Although protected by a solid dome of bone 10 in (25 cm) thick, one has become dizzy and is about to plummet to its death.

• THE END OF THE DINOSAURS •

Why Did They Vanish?

BIG BANG
According to one theory, several volcanic eruptions produced climatic changes that wiped out the dinosaurs.

METEORITE HITS
Perhaps a giant meteorite hit the Earth, causing dust clouds, acid rain, storms and huge waves.

END OF AN ERA
When the dinosaurs died out, all large land animals disappeared. Late Cretaceous mammals were small (*Alphadon*, shown here, was only 1 ft [30 cm] long) but evolved rapidly into thousands of new species to replace the dinosaurs.

The extinction of the dinosaurs 65 million years ago was the most mysterious and dramatic disappearance of a group of animals in the history of the Earth. But the dinosaurs were not the only animals to die out. More than half of the world's animals also disappeared, including the pterosaurs and large marine reptiles. The number of species of dinosaurs had been dropping for at least eight million years, but some species were common right up to the "K/T boundary," which marks the end of the Cretaceous Period and the beginning of the Tertiary Period. Some scientists believe that a volcanic disaster or a giant meteorite wiped out the dinosaurs. Others argue that such a disaster—causing disease, rising sea levels and gradual changes in climate—would have affected all animal life. Another theory combines these thoughts: changes in weather and sea levels had already reduced the amount of land and food for dinosaurs, and they were unable to cope with a sudden disaster.

Q: How do scientists explain the disappearance of the dinosaurs?

INTO THE FUTURE
Even today, habitat loss through earthquakes, storms or human activity such as clearing forests, is threatening the future of many animals.

STRANGE BUT TRUE
People have produced some weird and wonderful theories to explain why the dinosaurs disappeared. Some suggest they died of boredom, "drowned" in their own droppings, were hunted by aliens, or even committed suicide!

VICTIMS AND SURVIVORS
None of the current theories can explain why some animal groups disappeared, while others survived. Pterosaurs died out, but birds did not. Dinosaurs vanished, but small land reptiles and mammals survived. Mosaurs, plesiosaurs and pliosaurs were wiped out, but turtles and crocodiles are still alive today.

Victims	K/T Boundary	Survivors
Dinosaurs		
Pterosaurs		
Plesiosaurs		
Ammonites		
Mammals		
Crocodiles		
Lizards and snakes		
Turtles and tortoises		
Amphibians		
Fishes		
Insects		
Birds		

• THE END OF THE DINOSAURS •

Surviving Relatives

FEATHER FOSSILS
The detailed impressions of feathers on this *Archaeopteryx* fossil confirm an important evolutionary link between reptiles and birds.

SCALY SURVIVORS
Crocodilians have hardly changed since the beginning of the Cretaceous Period. They have evolved slowly because they live in a stable environment.

Dinosaurs are dead, but it seems that certain dinosaur features live on in other animals. Dinosaurs and birds, for example, are very different animals but they have many characteristics in common. Scientists are now convinced that dinosaurs were the ancestors of birds. The skeleton of *Archaeopteryx*, the earliest known bird, was very similar to that of the lizard-hipped carnivorous dinosaur *Compsognathus*. Many scientists now classify *Archaeopteryx* as a small, flesh-eating dinosaur with feathers (fossilized feather shown left). Dinosaurs are also related to crocodilians, which survived the great extinction at the end of the Cretaceous Period. Crocodilians and dinosaurs have very similar skulls and common ancestors—the archosaurs. The dinosaurs disappeared, but crocodilians today are almost the same as their ancestors. Their way of life has changed little in 150 million years.

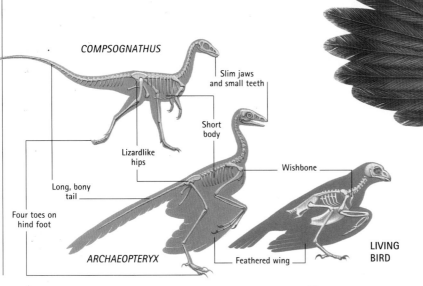

FROM DINOSAUR TO BIRD
Fossil records show a strong similarity between small carnivorous dinosaurs that ran upright on long, slim hind legs, *Archaeopteryx,* and living birds.

FLIGHT PATH
Archaeopteryx could fly by flapping its broad wings, but more often it would dive from its perch onto prey such as insects and small reptiles.

FAMILY TREE

- Alligators and crocodiles
- Ornithischian dinosaurs
- Saurischian dinosaurs
- Birds

Ornithosuchus

TRIASSIC | JURASSIC | CRETACEOUS | TERTIARY | QUATERNARY

RESEMBLING THE PAST

Some of today's birds are very similar to dinosaurs in their structure and behavior. The secretary bird of Africa rarely flies, but runs after insects, small reptiles and mammals on its hind legs—just as the dinosaur *Compsognathus* did. Baby hoatzins, from South America, use claws on the front of their wings to climb around in trees—just as *Archaeopteryx* did.

Hoatzin

Secretary bird

49

Allosaurus
Al-oh-sore-us:
"Different reptile"
Group: meat eater
Period: late Jurassic/early Cretaceous
Discovered: North America, 1877
Size: up to 39 ft (12 m)

Plateosaurus
Plat-ee-oh-sore-us:
"Flat reptile"
Group: plant eater
Period: late Triassic
Discovered: Europe, 1837
Size: 26 ft (8 m) long

Coelurus
Seel-ure-us:
"Hollow tail"
Group: meat eater
Period: late Jurassic
Discovered: North America, 1879
Size: 7 ft (2 m) long

Stegosaurus
Steg-oh-sore-us:
"Roof lizard"
Group: plant eater
Period: late Jurassic
Discovered: United States, 1877
Size: up to 30 ft (9 m)

Coelophysis
Seel-oh-fie-sis:
"Hollow shape"
Group: meat eater
Period: late Triassic
Discovered: North America, 1889
Size: 10 ft (3 m) long

Euoplocephalus
You-op-loh-seff-a-lus:
"True plated head"
Group: plant eater
Period: late Cretaceous
Discovered: North America, 1910
Size: 20 ft (6 m) long

Saltasaurus
Salt-a-sore-us:
"Salt reptile"
Group: plant eater
Period: late Cretaceous
Discovered: South America, 1970
Size: 39 ft (12 m) long

• THE END OF THE DINOSAURS •

Identification Parade

Dinosaurs marched across 150 million years, and represented an amazingly successful and varied group of land animals. They prospered for many times longer than human beings have. Although we will never know exactly how many kinds of dinosaurs there were, we know enough about their evolution to see their steady progress from only a few species during the Triassic Period to almost twice as many during the Jurassic Period. This was followed by an incredible "flowering" during the Cretaceous Period, when there were more species of dinosaurs than during both preceding periods. Dinosaurs have taught us many valuable lessons about evolution and about how a group of animals spread across the land, and dominated the world before disappearing. But we are left with only tantalizing clues about how dinosaurs lived.

Pachycephalosaurus
Pack-ee-seff-ah-low-sore-us:
"Thick-headed reptile"
Group: plant eater
Period: late Cretaceous
Discovered: North America, 1943
Size: up to 26 ft (8 m) long

Maiasaura
My-ah-sore-ah:
"Good mother lizard"
Group: plant eater
Period: late Cretaceous
Discovered: North America, 1979
Size: up to 30 ft (9 m)

Brachiosaurus
Brak-ee-oh-sore-us:
"Arm reptile"
Group: plant eater
Period: late Jurassic
Discovered: North America, 1903
Size: up to 75 ft (23 m)

Deinonychus
Die-non-i-kus:
"Terrible claw"
Group: meat eater
Period: early Cretaceous
Discovered: North America, 1969
Size: 10 ft (3 m) long

Hypsilophodon
Hip-sih-loh-foe-don:
"High-ridged tooth"
Group: plant eater
Period: early Cretaceous
Discovered: Europe, 1870
Size: 7 ft (2 m) long

Ouranosaurus
Oo-ran-oh-sore-us:
"Brave reptile"
Group: plant eater
Period: early Cretaceous
Discovered: Africa, 1976
Size: 23 ft (7 m) long

Dinosaurs Today

When Richard Owen invented the name dinosaur 150 years ago, we knew of just nine species. Today we know of at least 1,000, which includes an incredible variety of plant eaters, meat eaters and egg thieves, dinosaurs with horns and crests, spikes and razor-sharp claws. We are surrounded by dinosaurs: in museums, movies and theme parks, such as this display of robot dinosaurs in Japan. Even though they disappeared 65 million years ago, dinosaurs are "alive" in our imaginations. We are still learning about them; in fact, children today know more about dinosaurs than most adults do.

Tyrannosaurus
Tie-ran-oh-sore-us:
"Tyrant lizard"
Group: meat eater
Period: late Cretaceous
Discovered: North America, 1902
Size: up to 46 ft (14 m)

Parasaurolophus
Par-ah-sore-ol-oh-fus:
"Parallel-sided reptile"
Group: plant eater
Period: late Cretaceous
Discovered: North America, 1923
Size: 33 ft (10 m) long

Triceratops
Try-ser-ah-tops:
"Three-horned face"
Group: plant eater
Period: late Cretaceous
Discovered: North America, 1889
Size: up to 30 ft (9 m)

Struthiomimus
Strooth-ee-oh-mime-us:
"Ostrich-mimic"
Group: meat eater
Period: late Cretaceous
Discovered: North America, 1917
Size: up to 13 ft (4 m)

Under the Sea

- Why is the sea blue?
- What happens to salmon after they breed?
- What kinds of creatures live at the bottom of the ocean?

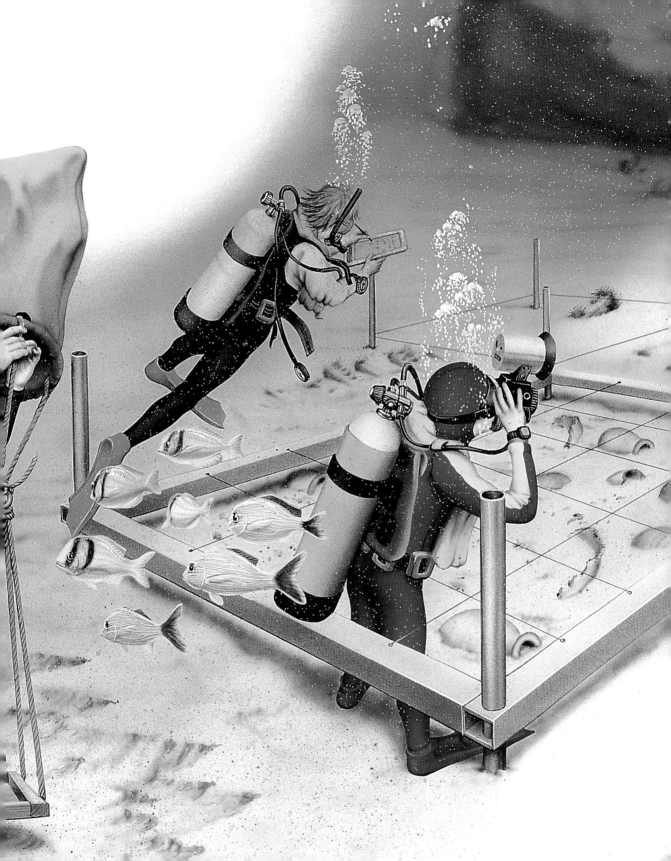

• UNDERWATER GEOGRAPHY •

Our Oceans

If you looked at the Earth from space, it would look extremely blue. This is because vast oceans cover almost two-thirds of our "blue planet." The Pacific, the Atlantic, the Indian, the Arctic and the Southern are the world's major oceans. They were formed by complex geological processes that continue to affect the Earth. The Earth is made up of seven main parts, called lithospheric plates, formed from the upper part of the Earth's mantle layer and the crust. Many millions of years ago, these parts all fit together. But nothing on Earth is fixed, and these plates are constantly moving (at about the same speed your fingernails grow) over a layer of soft, squishy rock called the asthenosphere that lies beneath the crust. When two plates move away from each other, hot melted rock, or magma, rises to fill the space and forms a new sea floor. In this way, ocean basins can grow gradually over millions of years. Five million years ago, the Red Sea was a shallow basin. Now, as the sea floor spreads, scientists think it has the makings of a new ocean.

THE BEGINNING
About 250 million years ago, there was one huge continent known as Pangaea, but before this, the real "beginning" is still shrouded in mystery.

PLATES MOVE APART
About 200 to 130 million years ago, Pangaea broke into separate pieces.

Did You Know?
Alfred Lothar Wegener, who lived between 1880 and 1930, was a German scientist. He was the first to suggest that many millions of years ago the world was one huge supercontinent.

Forces below
The core of the Earth is getting gradually hotter. When plates move apart, more magma bubbles up to the surface. As the channel of magma gets wider, it pushes against the sea floor, which buckles and forms ridges of crust. When this happens, the sea floor spreads outwards, pushing the areas of land further and further apart.

Red Sea Spreading

The African and Arabian plates began moving apart between five and ten million years ago. As this movement continues at a rate of about $1/2$ in (1 cm) a year, the basin of the Red Sea is spreading slowly. Astronaut Eugene Cernan photographed Africa and the Arabian Peninsula as *Apollo 17* traveled toward the moon in 1972. The gash you can see in the continental crust is called the Great Rift Valley. It runs from the Jordan Valley and Dead Sea in the north down through East Africa in the south, and was probably caused by the movement of the plates.

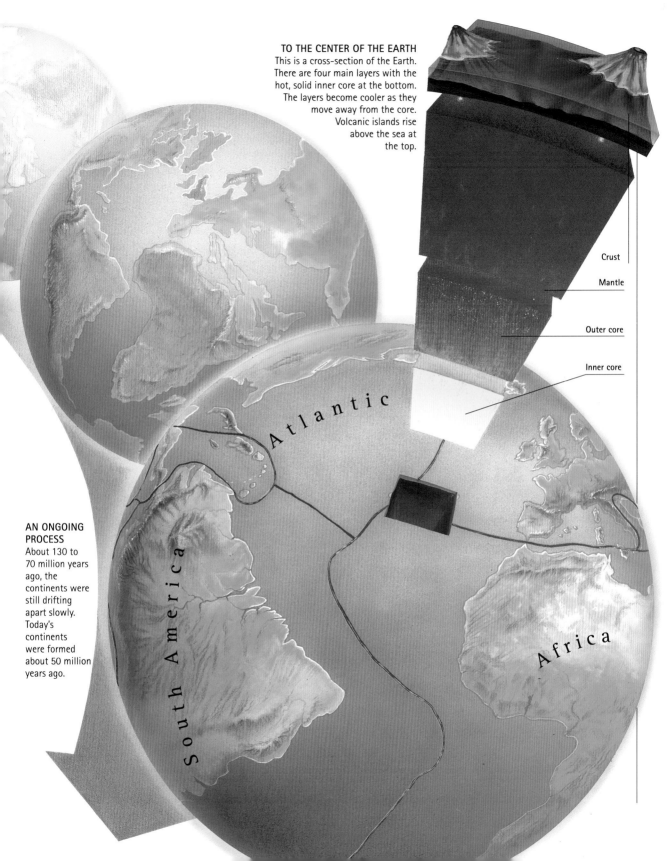

TO THE CENTER OF THE EARTH
This is a cross-section of the Earth. There are four main layers with the hot, solid inner core at the bottom. The layers become cooler as they move away from the core. Volcanic islands rise above the sea at the top.

Crust
Mantle
Outer core
Inner core

Atlantic

South America

Africa

AN ONGOING PROCESS
About 130 to 70 million years ago, the continents were still drifting apart slowly. Today's continents were formed about 50 million years ago.

• UNDERWATER GEOGRAPHY •

The Sea Floor

If all the water in the world's oceans was sucked away, we would be able to see the amazing landscape of the sea floor. With huge mountains and deep valleys, slopes and plains, trenches and ridges, it is surprisingly similar to the landscape of dry land. Modern ships and equipment have made it possible for us to learn about this hidden area. Between 1968 and 1975, the deep-sea drilling ship *Glomar Challenge* bored more than 400 holes in the sea bed and collected rock samples to be examined. These helped scientists piece together an accurate picture of the sea floor. They were able to detail its many features, such as a shallow continental shelf that extends from the land into the sea and may once have been dry land; and a continental slope, where the continent ends and the underwater land plunges to the very depths of the sea floor. Scientists continue to chart more of this underwater land with the help of computer images of underwater land forms and maps of the sea bed.

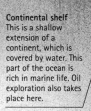

Continental shelf
This is a shallow extension of a continent, which is covered by water. This part of the ocean is rich in marine life. Oil exploration also takes place here.

Continental slope
This is the gently sloping, submerged land near the coast that forms the side of an ocean basin.

VOYAGE TO THE DEEP
This diver looks like an underwater astronaut as he dangles from a line attached to a vessel above.

LAYING CABLE
A diver and an underwater cable layer install telephone cables on the continental shelf.

RESEARCH INSTRUMENTS
Scientists collect and analyze deep-sea specimens and other information from the ocean floor to learn how the underwater landscape was formed. They use instruments such as the bathythermograph, which measures underwater temperatures, and the fisher scoop, which gathers up small samples of sand and mud from the sea bed.

Fisher scoop

Bathythermograph

PILLOW LAVA
When hot gases and liquid bubble up to the surface of the ocean floor, they harden and turn into lava. This is pillow lava, which has become part of the sea bed near the Galapagos Islands.

Seamounts
Most of these underwater volcanoes remain beneath the sea. Those that rise above the surface form islands.

Guyots
These are flat-topped seamounts.

Seeing with Sound

This map of the sea bed of the Great Barrier Reef in Australia shows a 19-mile (30-km) wide section of the seaward slope. It was produced by GLORIA, a mapping instrument that sends waves of sound energy down to the sea floor and records the returning echoes.

GLORIA
This instrument is attached to its "mother" ship by a conducting cable. It can reach depths of 164 ft (50 m) below the surface.

Abyssal plains
These are some of the flattest places on Earth. They spread out from the oceanic ridge to the edges of the continents.

Oceanic ridge
This is a ridge that rises when new sea floor wells up from inside the Earth.

Oceanic trench
A long, narrow valley, or trench, usually forms next to islands or beside coastal mountain ranges.

Did You Know?
Some of the world's deepest sea trenches extend further downwards than the highest mountains on land rise upwards.

• UNDERWATER GEOGRAPHY •

Sea Upheavals

The ocean is always moving. Its surface can change from calm and mirrorlike to wild and treacherous. Most waves at sea are caused by wind. The waves created by the gales that blow during a tropical cyclone are 46 ft (14 m) and higher. The largest wave known to have been caused by the wind was 112 ft (34 m) high. Waves can also be created by volcanic eruptions or earthquakes under the sea. These waves are known as tsunamis (pronounced soo-nah-mees). They are wide columns of water that reach down to the sea floor and can travel for great distances, at the speed of a jet plane, across the ocean. The surface of the ocean can also be changed by colliding currents. When the tide turns, the opposing currents meet and may create a whirlpool. One famous whirlpool is the fearsome Maelstrom off the coast of Norway. The thunder of its crashing eddies of water can be heard 3 miles (5 km) away.

Q: What causes changes in waves?

WHIRLING WINDS
A waterspout is a whirling column of air, laden with mist and spray. First cousin to the tornado, it can occur when rising warm, moist air meets cold, dry air. Sometimes schools of fish are sucked up by the fury of the spout, which can reach nearly 4 miles (6 km) into the air. Waterspouts rarely last more than 60 minutes, and while they are spectacular, they seldom cause any serious damage.

WALL OF WATER

People who live in coastal cities can be affected by sea upheavals. Imagine how frightening it would be to see an enormous wall of water rushing toward you. Your first reaction would be to run, but to where? The impact of the wave could destroy your whole city. Thousands of years ago a large part of Mauna Loa, one of the volcanic Hawaiian Islands, collapsed into the sea. This landslide produced a tsunami that traveled to the next island, Lanai, and crashed across it to a height of 918 feet (280 m). If such an event occurred today, all coastal areas in the Hawaiian Islands would be damaged. Waves of up to 98 ft (30 m) could roll into the city of Honolulu.

A DEVASTATING FORCE
A hurricane has a wind of force 12 or above on the Beaufort Scale, and it may be 400 miles (645 km) wide. This photograph of a hurricane called Elena was taken from the Space Shuttle *Discovery*.

THE BEAUFORT SCALE
This scale uses the numbers 1 to 12 to indicate the strength of wind at sea. At 0, the sea is as calm as a mirror; at 6 there is a strong breeze and large waves 10 ft (3 m) high. At 12, a hurricane is raging and the waves are more than 46 ft (14 m) high.

STORMING AWAY
This dramatically colored image of a severe storm in the Bering Sea was taken from a satellite in space.

Force 2

Force 8

Force 12

• UNDERWATER GEOGRAPHY •

Currents and Tides

Ocean currents are the massive bodies of water that travel long distances around the world. The major force that produces the currents is the wind. There are seven main ocean currents and thousands of smaller ones. They move in large, circular streams at about walking pace (1–5 knots). In the Northern Hemisphere, currents move in a clockwise direction; in the Southern Hemisphere they are counterclockwise. Winds carry the warm or cold water currents along the shorelines, affecting the climate of the various continents on the way. The Gulf Stream, for example, is a current that carries warm water from the Caribbean Sea, up the east coast of the United States and then to the west coasts of Britain and Northern Europe. Without the Gulf Stream, these areas would be much colder. Oceans are also influenced by the "pull" of the moon and the sun. This pull causes the tides. Each day the level of the sea rises and falls and then rises and falls again. Each high tide and the following low tide are about six hours apart. The difference in height between high tide and low tide is called the tidal range. The largest tidal ranges are found in bays and estuaries. The Bay of Fundy in Canada has a tidal range of 49 ft (15 m), the highest in the world. On open coasts the tidal range is usually 6–10 feet (2–3 m).

THE PULL OF THE MOON
As the moon is much closer to the Earth than the sun is, its pull is greater. The ocean waters on the side of the Earth facing the moon are pulled the most, resulting in a high tide. As the Earth itself is also pulled towards the moon, the waters on the other side of the planet form another, though slightly smaller, high tide.

Spring and Neap Tides

The highest and lowest tides occur when the Earth, the moon and the sun are in line with each other. These tides are called spring tides. When the sun and the moon form a right angle with the Earth, their combined pull is weakest and the difference between high and low tide, the tidal range, is at its lowest. These tides are called neap tides. Spring and neap tides occur twice a month.

Spring tide — Sun
Full/New moon

Neap tide — Half moon — Sun

IN FAR-FLUNG CORNERS
In 1977, Nigel Wace threw 20 wine bottles overboard from a ship traveling between South America and Antarctica to try to discover how far and how fast ocean litter travels. Most of the bottles took two years to drift to Western Australia and nearly three years to reach New Zealand. Others reached southern Africa, the Seychelles and Easter Island. Because there is so much litter in the oceans, Wace says that today he would not throw any trash into the sea, even for the sake of an experiment.

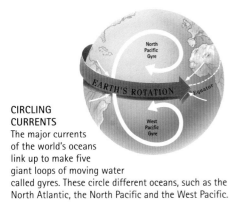

CIRCLING CURRENTS
The major currents of the world's oceans link up to make five giant loops of moving water called gyres. These circle different oceans, such as the North Atlantic, the North Pacific and the West Pacific.

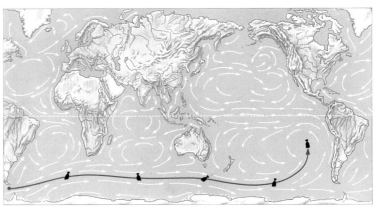

MOVING OCEAN CURRENTS
In the tropics, strong winds push currents towards the equator. In the northern and southern seas, westerly winds push currents eastward. When they reach a continent, they change direction. The spin of the Earth also influences the direction of the currents: those in the northern part of the world are pushed to the right, while those in the southern part are pushed to the left. This phenomenon is called the Coriolis effect.

Strange but True
When 80,000 Nike shoes were swept into the sea from a ship traveling between South Korea and Seattle, Curtis Ebbesmeyer traced their path to learn about ocean currents. The shoes began to wash up on the west coast of the United States about a year later.

• LIFE IN THE SEA •
The Seashore

The land meets the sea at the seashore, which is the home of many animals. Hundreds of species of crabs patrol sandy seashores and hide in rockpools, searching for scraps of food. Crustaceans or mollusks have shells or other hard casing to protect them from birds, the hot sun and the pounding waves. Sand hoppers feed on rotting plants, especially seaweed that has been washed up onto the beach. Sea urchins graze on tiny animals and plants from rocks; and starfish feed on coral and shellfish. Certain corals provide a safe shelter for other seashore animals. Some fish have also adapted to life near the shore. The weeverfish hides in the sand, ready to eat any small fish or crabs that swim nearby. Razor clams and burrowing sea anemones disappear into the sand when they have caught their prey.

Drawn in
Anemones are anchored in one place by their stalks. Their tentacles shorten when fish swim into them and pull the prey into the open mouth of the anemone.

INSIDE A STARFISH

The round center of a starfish body holds the stomach. The anus is above the stomach and the mouth is below. Canals holding water, branches of nerves and intestines spread into each of the five arms. If an arm is broken off, a starfish can grow a new one in a few weeks. A starfish has tubes inside its body that pump water in and out of its many tube feet. As the water pressure builds up, the feet become longer and they bend. This action propels the starfish along. Each tube foot has a little sucker on the end, which the starfish uses to climb rocks and to open shellfish.

Water enters here

Tubes pumping water

Tube feet

Garibaldi
Bright orange male garibaldi seek out small crevices or overhangs in their rock-reef homes. Female garibaldi spawn with males that hold the best nest sites.

Acorn barnacles
The rocky seashore is home to many acorn barnacles. They can feed only when the tide comes in and they are submerged.

Sea otters
These live on the shores of the northern Pacific Ocean. Sea otters use their sharp teeth and strong front paws to crack open the hard shells of crabs.

Kelp
This is a type of large, brown seaweed. It provides food and shelter for all kinds of sea creatures.

Mulberry whelk
This feeds on dead or dying animals.

Octopus
This has sharp eyesight and a large brain.

Periwinkles
These rough periwinkles can be found just below the waves on rocky shores.

Suit of armor
The chiton's shell has eight plates that fit one against the other.

Sea urchins
These use their long, sharp spines to defend themselves. Sometimes these spines contain a painful venom.

• LIFE IN THE SEA •
Coastal Seas

The coastal seas are the richest areas of the ocean. They teem with sea life and are very popular for fishing and trawling. Most of the fish and shellfish we eat are caught in these shallow waters, which are down to 200 ft (60 m) deep. They spread out over the outer parts of continents and larger islands. Coastal sea water is alive with plankton, tiny drifting plants and animals. When blue and yellow light waves bounce back off this plankton, it makes the water look very green. Fast-swimming fish, such as yellowtails, bluefish, striped bass and some types of tuna, feed off smaller mackerels, sardines and herrings close to the coasts. Humpback whales give birth in warm coastal seas, then push their newborn calves to the surface for their first breath.

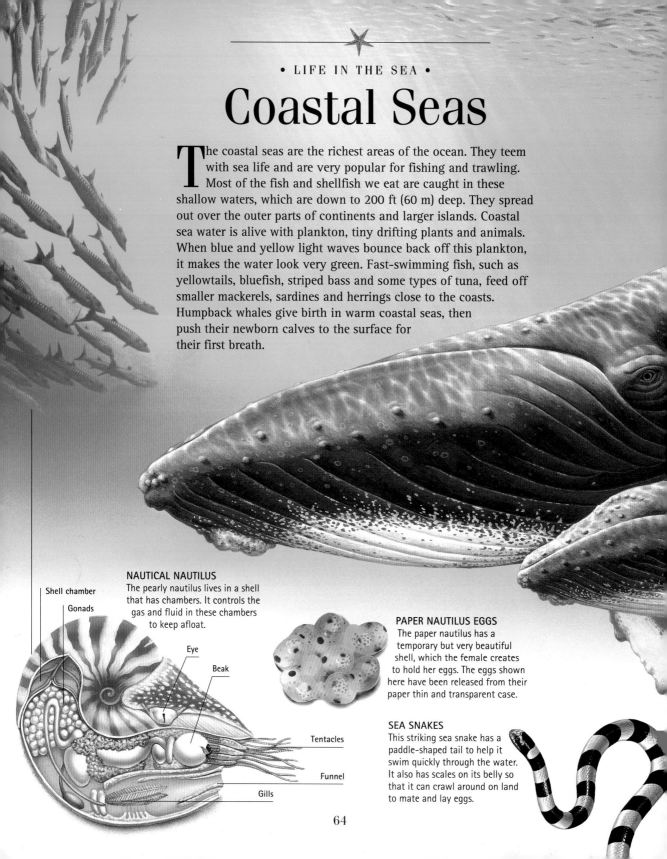

NAUTICAL NAUTILUS
The pearly nautilus lives in a shell that has chambers. It controls the gas and fluid in these chambers to keep afloat.

PAPER NAUTILUS EGGS
The paper nautilus has a temporary but very beautiful shell, which the female creates to hold her eggs. The eggs shown here have been released from their paper thin and transparent case.

SEA SNAKES
This striking sea snake has a paddle-shaped tail to help it swim quickly through the water. It also has scales on its belly so that it can crawl around on land to mate and lay eggs.

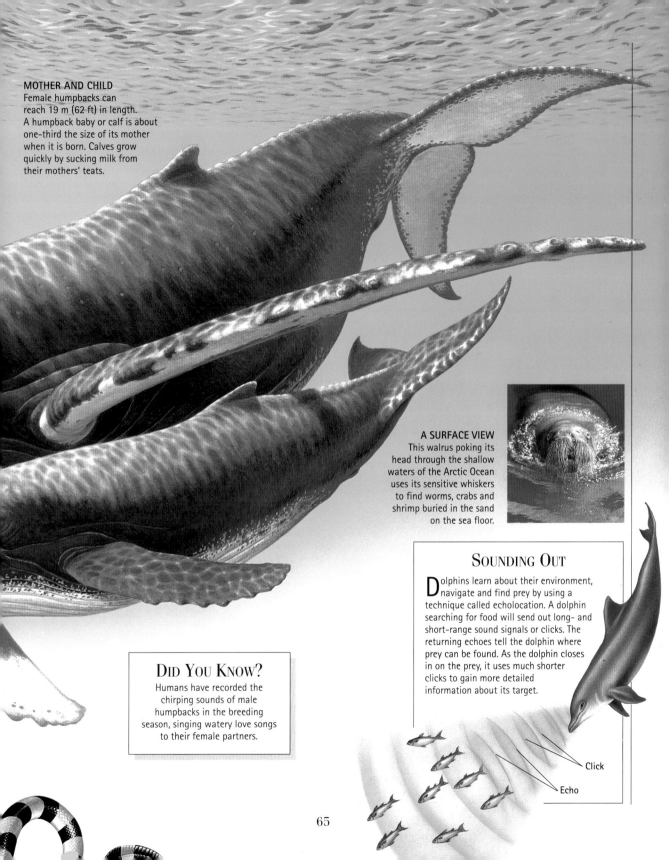

MOTHER AND CHILD
Female humpbacks can reach 19 m (62 ft) in length. A humpback baby or calf is about one-third the size of its mother when it is born. Calves grow quickly by sucking milk from their mothers' teats.

A SURFACE VIEW
This walrus poking its head through the shallow waters of the Arctic Ocean uses its sensitive whiskers to find worms, crabs and shrimp buried in the sand on the sea floor.

SOUNDING OUT

Dolphins learn about their environment, navigate and find prey by using a technique called echolocation. A dolphin searching for food will send out long- and short-range sound signals or clicks. The returning echoes tell the dolphin where prey can be found. As the dolphin closes in on the prey, it uses much shorter clicks to gain more detailed information about its target.

Click
Echo

DID YOU KNOW?

Humans have recorded the chirping sounds of male humpbacks in the breeding season, singing watery love songs to their female partners.

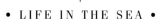

• LIFE IN THE SEA •

Coral Reefs

Brightly colored fish and thousands of other sea creatures live in the shelter of coral reefs. These marine homes grow in warm shallow seas and are built by coral animals, or polyps, with soft bodies and mouths that are ringed by stinging tentacles. The polyps construct thimble-shaped skeletons of limestone around themselves. As the polyps grow upwards, they keep dividing in two. They leave their skeletons behind them, however, and these fuse together to make a coral reef. A living mass of growing polyps always forms a film of flesh on top of the skeletons. Each polyp has many round plant cells living in its body and these cells make food from a combination of sunlight, water and carbon dioxide (a process called photosynthesis). The corals are able to catch their own food with their stinging tentacles, but most of the food they eat is made by the plant cells. Coral reefs need the food from these plant cells to grow quickly.

HIDING OUT
A clownfish lives and hides from its enemies within the tentacles of the coral reef anemone. It escapes being stung by covering itself in a layer of mucus from the anemone. The coral is fooled into thinking that the fish is part of itself.

CORAL WATCHING
Coral reefs, such as this one at Taveuni Island in the Pacific Ocean, attract snorkelers and divers from all over the world. But coral reefs are very fragile, and some are being damaged by human contact.

CORAL COMMUNITIES
Many species of coral, such as sea fan coral, hard brain coral, bubble coral and soft fire coral, grow together. They live side by side with goldfish, giant clams, surgeonfish and many other sea dwellers.

CORAL SPAWNING
When coral spawn, some release their eggs and sperm to be fertilized in the water; others release sperm to fertilize the egg inside the polyp.

CORAL POLYPS
These tiny coral animals form coral colonies of different shapes and colors. Plant cells live within the tissues of most corals and these help the coral polyp to produce its limestone skeleton.

Tentacles

Mouth

CROWDED HOUSE
Crustaceans, fish, sea urchins, mollusks and clams are some of the many creatures that live on a coral reef.

IN QUICK PURSUIT
Killer whales are fast swimmers. They have cone-shaped teeth to catch and chew fish and smaller mammals.

• LIFE IN THE SEA •

Ocean Meadows

The ocean is like a giant meadow, providing food for all its creatures. The food web that operates under water is a complex system where large creatures prey on smaller creatures. Killer whales eat seals and sealions, which feed on fish and squid. Salmon enjoy small fish, which eat plankton— the tiny plants (phytoplankton) and animals (zooplankton) that float in the sunlight of the surface waters. Plankton is the basic source of food for ocean animals, and plants are the most important link in the food web. They use water, carbon dioxide and energy from sunlight to make plant food. If links in the food web are ever lost, others will take their place. Sardines once played a vital role in the food web off the coast of California. But they became scarce when too many of them were fished, and anchovies took their place.

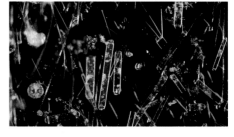

PHYTOPLANKTON
The sunlit, upper layer of the ocean teems with microscopic life, such as plant plankton, the basic food of the sea.

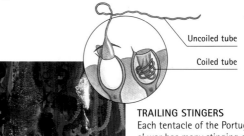

TRAILING STINGERS
Each tentacle of the Portuguese man o' war has many stinging cells and a sac containing a coiled, barbed tube. When a fish touches the cell, the tube uncoils, pierces the skin and delivers the venomous poison.

ZOOPLANKTON
Many kinds of microscopic animal plankton swim in the ocean. Some are the larvae of fish, which have just hatched, while others are small crustaceans, such as shrimp.

SEALIONS AND SEALS
These are the natural prey of large meat-eating mammals and fish, such as whales and sharks.

PREYING ON SMALL FISH
With mouth open and sharp teeth ready, a salmon swims after a school of herring.

THE WORLD OF PLANKTON
This satellite image of the world's oceans shows where plant plankton are common. Some areas have more plankton than others and the colors on the map indicate these from the most (red), through yellow, green and blue, to the least (violet). The grey parts show where there are gaps in the information collected.

TINY LINKS
Krill are small, shrimplike creatures, which live in vast numbers in the Southern Ocean. Whales, seals and seabirds all feed on krill, and they are an important link in the food web in this part of the world.

GIANT RAY
Not all large animals in the food web eat large prey. The manta ray glides through the water, mouth agape, sieving out tiny fish and crustaceans.

TORPEDO POWER
A squid speeds through the water, its torpedo-shaped body shooting out a water jet behind it, and grabs a fish with the special suckers on the ends of its tentacles. As it is a very fast swimmer, a squid can catch prey easily.

FILTERING THROUGH
Herring feed on plankton, which they filter from the water.

Q: What is the most important link in the food web?

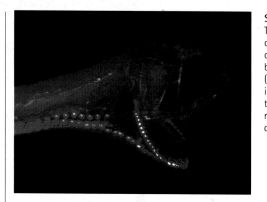

SLOANE'S VIPER
This viperfish has light organs, or photophores on its belly. Despite being only about 1 ft (30 cm) long, it has impressive jaws and teeth. It is one of the most feared of the deep-sea predators.

SWALLOWER
This fish has a hugely expandable stomach and is known as a swallower. It is able to eat fish that are longer and larger than itself.

• LIFE IN THE SEA •

Life in the Twilight Zone

Imagine the world at dusk. It is hard to see in the gloom, and shapes blur into the blackness. This is the atmosphere of the twilight zone, 656–3,280 ft (200–1,000 m) below the sunlit surface of the sea. Only blue light remains in this cold, deep-sea zone, the home of many interesting creatures that have adapted to life in this part of the ocean. Regalecus, the king of the herring, shares this murky world with lampris, a silver-spotted fish. Giant squid, which sometimes rise to the surface at night, loom in the depths with swordfish and big-eye tuna. Many of the fish in the twilight zone glow in the dark. They have bacteria that produces light—a process called bioluminescence. These animals use bioluminescence in different ways: some send out light patterns to attract mates in the darkness; several kinds of fish have bioluminescent organs on the lower half of their bodies, which they use for camouflage; others temporarily blind their predators with sudden flashes of light.

DID YOU KNOW?
Some kinds of fish and shrimp use bioluminescence to camouflage themselves. They have light organs on the lower half of their bodies that they use to blend in with light filtering from the surface. When predators look upwards, they cannot see the shape of their prey.

MOLA MOLA
This ocean sunfish has a very distinctive body shape and can be up to 10 ft (3 m) long.

KING-OF-THE-SALMON
Native Americans call the ribbonfish king-of-the-salmon. They believe it leads Pacific salmon back to the rivers to spawn when the breeding season begins.

ANGLERFISH
The female anglerfish has a luminous lure. The bulblike bait on her head contains luminescent bacteria. This attracts prey to the anglerfish, which saves energy by not having to hunt for food.

LANTERNFISH
There are huge numbers of these fish in the deep sea. They are called lanternfish because they have light organs on their heads and bodies.

COLONIAL SEA SQUIRT
The sacklike body of the sea squirt has openings through which water enters and leaves.

FLASHLIGHT FISH
This fish can be seen from a distance of 98 ft (30 m) in the dark depths of the ocean.

SQUID
Many squid live in the ocean depths. They have well-developed senses and can propel themselves quickly through the water.

VIPERFISH
The curving fangs of the small viperfish make it a dangerous predator.

SEEING IN THE DARK
Flashlight fish are found in caves at the bottom of coral reefs. They have large light organs under their eyes that contain luminous, or glowing, bacteria. The fish use these light organs to feed, and to communicate with other flashlight fish. But glowing in the dark can create problems when trying to avoid predators. The flashlight fish is able to cover the light organ with a screen of pigmented tissue, called a melanphore. This means it can turn the light on and off— just like a flashlight.

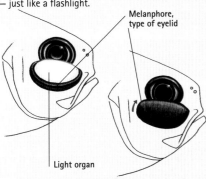

Melanphore, type of eyelid

Light organ

HATCHETFISH
These fish have light organs underneath their bodies that confuse predators swimming beneath them. Their eyes have large lenses that help them see small, glowing fish and shellfish.

SCALY DRAGONFISH
The thick, jellylike layer that covers the scales of this fish contains light organs.

• LIFE IN THE SEA •

Life on the Ocean Floor

Seeing in the dark
The US Navy submersible *Alvin* can carry its two crew members to a depth of 13,000 feet (3,960 m).

The ocean floor is cold, dark and still. The temperature never rises to more than just above freezing, and there is no light. This means there is little food, for plants cannot grow without energy from the sun. Deep-sea dwellers filter, sieve and sift the water and mud on the ocean floor to find tiny pieces of food that have dropped from the surface of the sea. These creatures of the deep have adapted well to their demanding environment. Some have soft, squishy bodies and large heads. They do not need strong skins and bones because there are no waves in this part of the ocean. Many are blind and move slowly through the water. Gigantic sea spiders, gutless worms and glass rope sponges are some of the unusual creatures that live in the inky blackness of the ocean floor.

Eelpout
These long fish live near underwater vents and eat tube worms.

Muscling in
Mussels and giant clams live on bacteria inside their bodies.

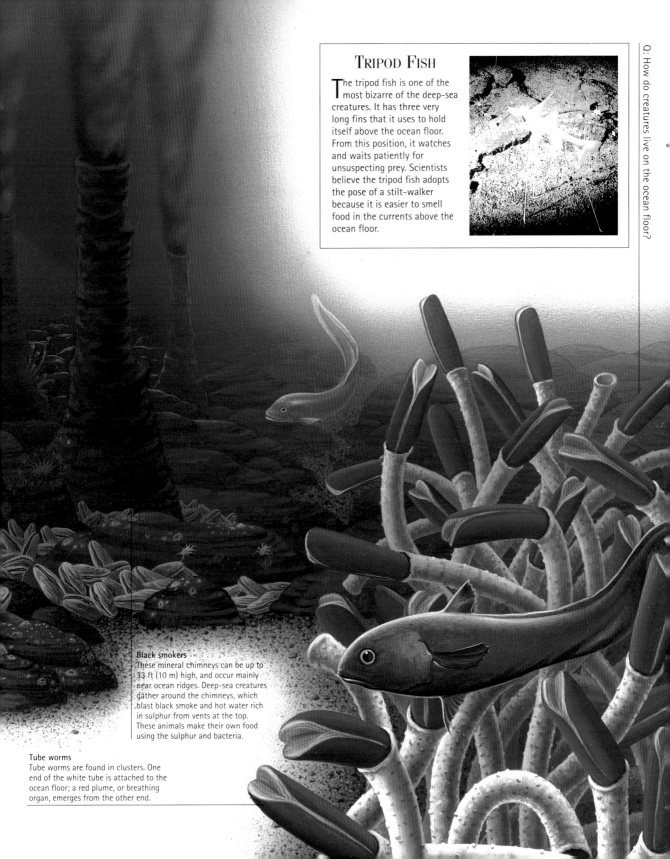

TRIPOD FISH

The tripod fish is one of the most bizarre of the deep-sea creatures. It has three very long fins that it uses to hold itself above the ocean floor. From this position, it watches and waits patiently for unsuspecting prey. Scientists believe the tripod fish adopts the pose of a stilt-walker because it is easier to smell food in the currents above the ocean floor.

Q: How do creatures live on the ocean floor?

Black smokers
These mineral chimneys can be up to 33 ft (10 m) high, and occur mainly near ocean ridges. Deep-sea creatures gather around the chimneys, which blast black smoke and hot water rich in sulphur from vents at the top. These animals make their own food using the sulphur and bacteria.

Tube worms
Tube worms are found in clusters. One end of the white tube is attached to the ocean floor; a red plume, or breathing organ, emerges from the other end.

• EXPLORING THE OCEANS •

Submersibles

FROM THE INSIDE
The control panels at the nerve centre of the submersible *Alvin* look as complex as those on any jumbo jet or spacecraft. The crew member uses radio-controlled headphones to communicate with the "mother" ship.

The ocean floor is many miles below the surface. While the safe maximum depth for scuba diving is 165 ft (50 m), the deepest parts of the ocean may be 7 miles (11 km) below the surface. The only way to reach such depths is by a submersible, a small submarine that dives from its "mother" ship. The United States submersible *Alvin* and the French vessel *Nautile* have visited the underwater site of the sunken ocean liner the *Titanic*. The crew of *Alvin* used an even smaller robot submersible, *Jason Jr.*, to probe inside places too small or too dangerous for *Alvin* to go itself. Larger submersible structures have also been used for research. The Hydrolab was launched in Florida in 1968, and for 18 years was the underwater home of scientists who observed and recorded the habits and behavior of lobsters, snapper, grouper and the hundreds of creatures living on a coral reef.

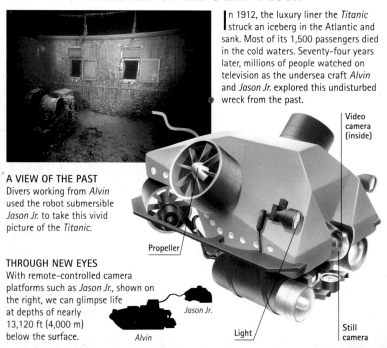

DISCOVERY ON THE OCEAN FLOOR

In 1912, the luxury liner the *Titanic* struck an iceberg in the Atlantic and sank. Most of its 1,500 passengers died in the cold waters. Seventy-four years later, millions of people watched on television as the undersea craft *Alvin* and *Jason Jr.* explored this undisturbed wreck from the past.

A VIEW OF THE PAST
Divers working from *Alvin* used the robot submersible *Jason Jr.* to take this vivid picture of the *Titanic*.

THROUGH NEW EYES
With remote-controlled camera platforms such as *Jason Jr.*, shown on the right, we can glimpse life at depths of nearly 13,120 ft (4,000 m) below the surface.

Video camera (inside)
Propeller
Jason Jr.
Alvin
Light
Still camera

74

THE NEW FRONTIER

The depths of the ocean are mysterious and still. We have been trying to explore them with metal helmets, diving suits, scuba gear and diving bells since the 1600s. With the help of advancing technology and modern submersibles, we are discovering more about this watery frontier.

How Deep Can They Go?

AQUALUNG
1943
165 ft
(50 m)

COUSTEAU'S DIVING SAUCER
1959
1,350 ft
(410 m)

JIM
1971
2,000 ft
(610 m)

NR-1
1969
2,300 ft
(700 m)

BATHYSPHERE
1934
3,028 ft
(925 m)

DSRV-1
1965
5,000 ft
(1,525 m)

CYANA
1959
9,800 ft
(2,990 m)

ALVIN
1964
13,000 ft
(3,960 m)

TRIESTE
1953
35,800 ft
(10,920 m)

ARCHIMÈDE
1962
36,000 ft
(10,980 m)

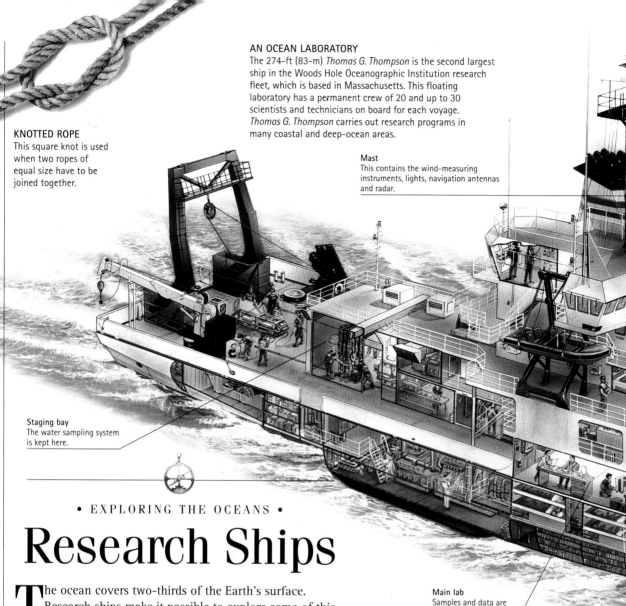

AN OCEAN LABORATORY
The 274-ft (83-m) *Thomas G. Thompson* is the second largest ship in the Woods Hole Oceanographic Institution research fleet, which is based in Massachusetts. This floating laboratory has a permanent crew of 20 and up to 30 scientists and technicians on board for each voyage. *Thomas G. Thompson* carries out research programs in many coastal and deep-ocean areas.

KNOTTED ROPE
This square knot is used when two ropes of equal size have to be joined together.

Mast
This contains the wind-measuring instruments, lights, navigation antennas and radar.

Staging bay
The water sampling system is kept here.

• EXPLORING THE OCEANS •

Research Ships

The ocean covers two-thirds of the Earth's surface. Research ships make it possible to explore some of this enormous area. They are specially equipped so that scientists can study deep currents and the structure of the ocean floor; learn how the ocean interacts with the Earth's atmosphere and how this affects climate and weather; and how natural and human disturbances, such as burning fossil fuels and releasing carbon dioxide into the air, affect the oceans. These vessels carry sophisticated navigation and communications systems, cranes and winches for their sampling and measuring devices, special mooring cables and tanks for live specimens. The requirements of their on-board laboratories often change from voyage to voyage. The ships stay in touch with their home ports by satellite.

Main lab
Samples and data are analyzed around the clock in the main lab.

WATER SAMPLERS
These scientists are preparing to send water sample bottles to the bottom of the ocean. Valuable information can be obtained by analyzing these samples.

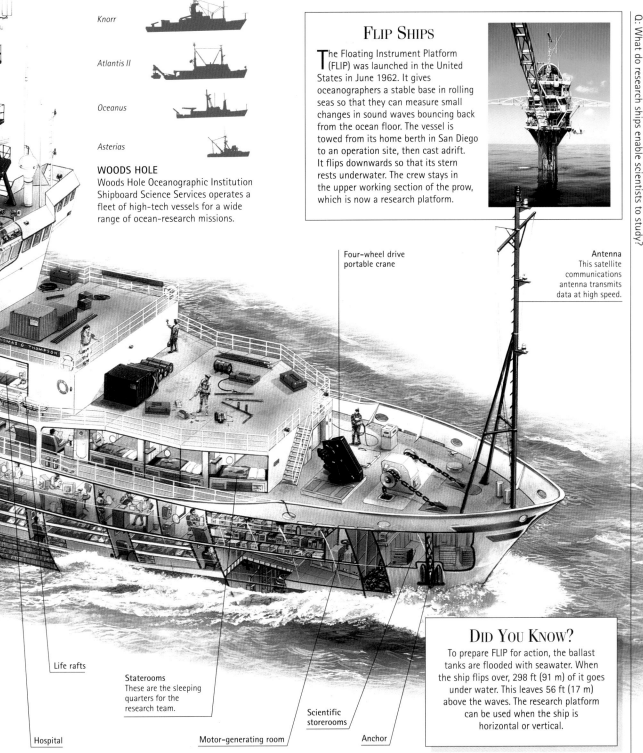

Knorr

Atlantis II

Oceanus

Asterias

WOODS HOLE
Woods Hole Oceanographic Institution Shipboard Science Services operates a fleet of high-tech vessels for a wide range of ocean-research missions.

FLIP SHIPS
The Floating Instrument Platform (FLIP) was launched in the United States in June 1962. It gives oceanographers a stable base in rolling seas so that they can measure small changes in sound waves bouncing back from the ocean floor. The vessel is towed from its home berth in San Diego to an operation site, then cast adrift. It flips downwards so that its stern rests underwater. The crew stays in the upper working section of the prow, which is now a research platform.

Four-wheel drive portable crane

Antenna
This satellite communications antenna transmits data at high speed.

Life rafts

Hospital

Staterooms
These are the sleeping quarters for the research team.

Motor-generating room

Scientific storerooms

Anchor

DID YOU KNOW?
To prepare FLIP for action, the ballast tanks are flooded with seawater. When the ship flips over, 298 ft (91 m) of it goes under water. This leaves 56 ft (17 m) above the waves. The research platform can be used when the ship is horizontal or vertical.

Q: What do research ships enable scientists to study?

• OCEAN MYSTERIES •

Sea Legends

Early seafarers and explorers, searching for new lands, faced daily perils in unknown seas. They braved storms, icebergs, fog, hidden reefs and the unsettling calm, waiting for a flurry of wind to catch the sails. Sailors told of huge sea monsters; of mermaids and mermen; and of Neptune, the fiery god of the sea. Rumors and exaggerated tales of true and imagined sea creatures were exchanged at every port. Cartographers, drawing detailed maps of new routes and countries, even included pictures of dragon-like monsters roaring their way around the world's oceans. Many nations have legends about the sea. The ancient Greeks told stories of sirens (part woman, part bird) whose sweet songs lured mariners to their death on jagged rocks. Ulysses, one of the heroes of Greek mythology, had to put wax in his sailors' ears to stop them from jumping into the sea as they sailed past the sirens' island. He lashed himself to the mast so that he could hear their singing, but not be charmed by it.

SEA MONSTERS
Do they really exist? Or are they exaggerated versions of real sea giants, such as sawfish, narwhals or humpback whales? The octopus-like sea monster wrapping itself around this ship is said to be a kraken, a mythical creature that appears off the shores of Norway. It must seem very real to the sailors clinging desperately to the ropes on this sinking ship.

WOMEN OF THE WAVES
Mermaids often appear in legends, with long flowing tresses of hair, decorated with delicate shell combs. Stories tell of mermaids enticing humans into the sea, and drowning them in its depths.

Q: How do legends begin?

THE LOCH NESS MONSTER
The sea is not the only water surrounded by mysteries and myths. The deep lake called Loch Ness in northern Scotland is a strange and lonely place, often shrouded by mist. Visitors claim to have seen and photographed a monster rising silently above the surface, and disappearing mysteriously. Scientists have investigated the sightings but have never found the monster. Nor have they proved that it does not exist. If you visit the lake, you can see a video of the elusive creature "Nessie."

SEA GOD
The Romans believed that Neptune was the god of the sea. He ruled the many creatures that lived below the waves.

MONSTER ACT
This photograph is said to prove once and for all that the Loch Ness Monster really exists. But is this dark shape really a monster? Could it be a whale or a mystery submarine?

SHOWTIME FOR NESSIE!

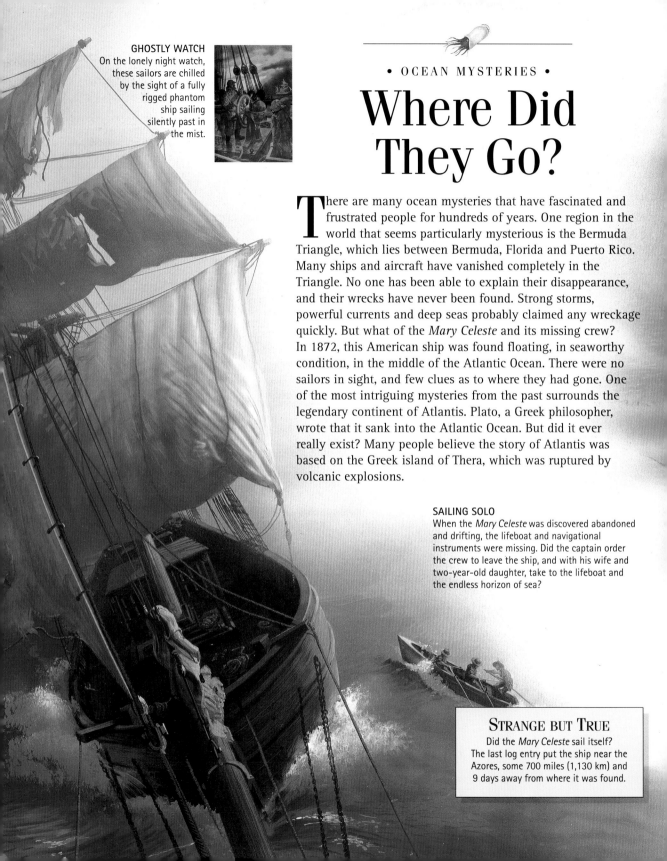

GHOSTLY WATCH
On the lonely night watch, these sailors are chilled by the sight of a fully rigged phantom ship sailing silently past in the mist.

• OCEAN MYSTERIES •

Where Did They Go?

There are many ocean mysteries that have fascinated and frustrated people for hundreds of years. One region in the world that seems particularly mysterious is the Bermuda Triangle, which lies between Bermuda, Florida and Puerto Rico. Many ships and aircraft have vanished completely in the Triangle. No one has been able to explain their disappearance, and their wrecks have never been found. Strong storms, powerful currents and deep seas probably claimed any wreckage quickly. But what of the *Mary Celeste* and its missing crew? In 1872, this American ship was found floating, in seaworthy condition, in the middle of the Atlantic Ocean. There were no sailors in sight, and few clues as to where they had gone. One of the most intriguing mysteries from the past surrounds the legendary continent of Atlantis. Plato, a Greek philosopher, wrote that it sank into the Atlantic Ocean. But did it ever really exist? Many people believe the story of Atlantis was based on the Greek island of Thera, which was ruptured by volcanic explosions.

SAILING SOLO
When the *Mary Celeste* was discovered abandoned and drifting, the lifeboat and navigational instruments were missing. Did the captain order the crew to leave the ship, and with his wife and two-year-old daughter, take to the lifeboat and the endless horizon of sea?

STRANGE BUT TRUE
Did the *Mary Celeste* sail itself? The last log entry put the ship near the Azores, some 700 miles (1,130 km) and 9 days away from where it was found.

FLIGHT 19'S LAST MISSION
On 5 December 1945, the sky droned with the engine noises of five torpedo bombers on a training flight from Florida. But flying across the Bermuda Triangle, the whole squadron vanished. During the last radio contact with their base, they said they were low on fuel and might have to land in the water. Rescue crews scoured the ocean for five days. They discovered no trace of the missing men or planes.

THE LOST KINGDOM

The Greek philosopher Plato was the first to write about the lost civilization of Atlantis. He said that thousands of years ago there was a large island in the Atlantic Ocean. The temples were decorated with gold, silver, copper and ivory; the people were very wealthy and lived in magnificent buildings. But, according to Plato, the people became greedy and dishonest, and the gods decided to punish them. During one day and night, violent eruptions shook the island and it disappeared, forever, into the sea.

Plato

MAPPED OUT
A map from the seventeenth century shows Atlantis as a very large island, midway between America and the Pillars of Hercules, at the entrance to the Mediterranean Sea.

BABY TURTLES
These turtles make an instinctive dash for the sea after hatching in the sand. But once in the sea, they are easy targets for predators, such as sharks. Most do not survive.

• OCEAN MYSTERIES •

Mysteries of Migration

Many of the animals in the world make long journeys, or migrations, each year. They move to warmer climates, to find food, or a safe place to breed and raise their young. Migrations can cover thousands of miles. Many polar seabirds migrate enormous distances, but the Arctic tern makes the longest journey of all creatures. Each year, it travels from the top to the bottom of the globe and back again— a journey of 9,300 miles (15,000 km). Whales mate and give birth in warm seas, but they migrate to polar seas to eat the huge amounts of krill they need. Marine turtles can spend more than a year building up the fat reserves they will need when they leave their feeding grounds. They journey across vast oceans to certain regions where they mate and lay their eggs. But how do marine turtles navigate over the open ocean with such accuracy? There is much about animal migration that continues to baffle scientists.

THE LIFE CYCLE OF A SALMON

Salmon lay their eggs in freshwater rivers and streams. Young salmon, called alevin, hatch in gravel on the river bed and remain there for several weeks. Then they begin to swim downstream to the salty ocean, where they will feed on fish, squid and krill. This migration usually takes place at night to avoid predators. Salmon spend up to four years at sea before returning to breed in the river in which they were hatched. Some adult salmon will travel thousands of miles to reach these rivers. After breeding, the salmon die.

SALMON EGGS
Salmon hide their large yolky eggs in the gravel of river beds to keep them safe from predators. Young salmon feed on their yolk sac.

SWIMMING UPSTREAM
Leaping sockeye salmon fight their way back up the Adams River, Canada, to their home spawning grounds.

ON THE MOVE
The larvae of European eels hatch in the Sargasso Sea, in the north Atlantic Ocean. Then they swim to the mouths of freshwater rivers and streams in North America and Europe, taking two to three years to make the journey. Here they change into elvers and gradually mature into adult eels.

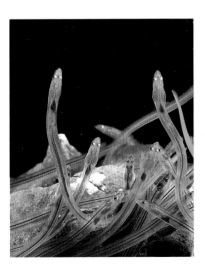

LOBSTER LINKS
When spiny lobsters migrate in the autumn, they form lines and march in single file across the ocean floor. Each creature stays in contact with the one in front. If an enemy appears, the lobsters back away from it and point their spiny antennae in an attack position.

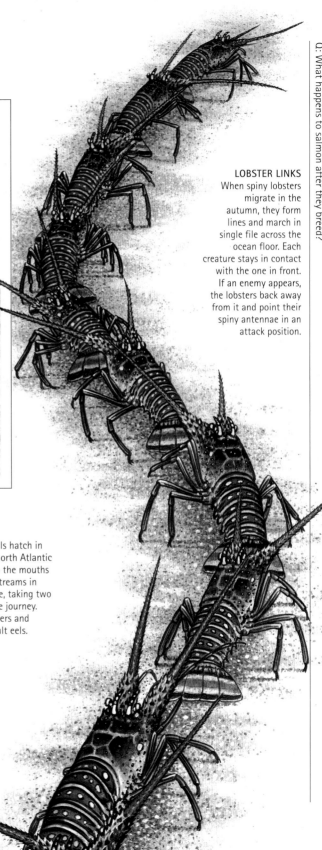

Q: What happens to salmon after they breed?

• EXPLOITING THE OCEANS •

Oils and Minerals

The most sought-after resources found in today's oceans are oil and natural gas. They are located offshore in many parts of the world, and new fields are being discovered all the time. Oil-rich countries sell to oil-hungry countries, and the trade in oil and gas affects the economy of the whole world. As it takes thousands of years for oil, a fossil fuel, to develop, there is a limited amount of oil in the world. When it is all gone, we will have to find other sources of energy, such as solar energy from the sun. Many useful minerals also come from the sea. Sea water is very salty and humans have extracted salt from the sea for thousands of years. Marine-based minerals form crusts, which cover parts of the ocean bottom.

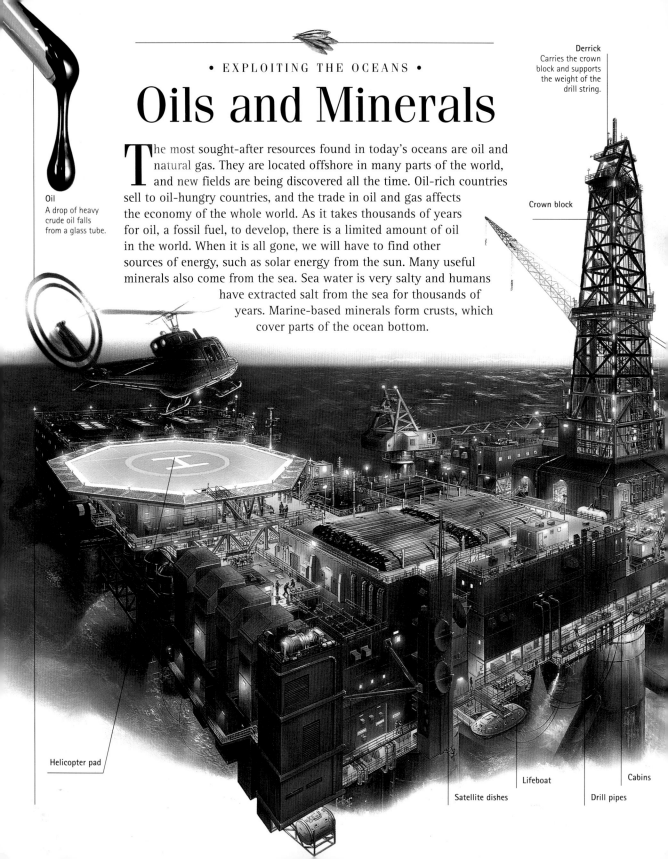

Oil
A drop of heavy crude oil falls from a glass tube.

Derrick
Carries the crown block and supports the weight of the drill string.

Crown block

Helicopter pad

Satellite dishes

Lifeboat

Drill pipes

Cabins

WHAT IS OIL?

Oil was formed when dead plants and animals sank to the bottom of the ocean. The dead matter was buried under piles of mud and sand, which turned into rock after millions of years. The decaying matter was squashed and, as the temperature and pressure increased, it collected in the sedimentary rock as droplets of oil held between the rock grains— just like a sponge holds water.

MINING FOR MANGANESE

Manganese is a hard, brittle metal element. It is used to make alloys, such as steel, harder and stronger. Manganese nodules are found on parts of the ocean floor. The ones shown here were dredged from the Blake Plateau in the North Atlantic Ocean, 1,378 ft (420 m) below the surface. Manganese nodules form in places where sediment builds up slowly. As each new layer of metal is added, the nodules grow bigger, sometimes joining together with other nodules. Most nodules look like small black potatoes and grow at the rate of a few millimeters every million years. Some of the world's mining companies have located rich deposits of deep-sea manganese nodules. However, mining manganese is an extremely expensive process. First, the nodules must be dredged, often in very deep waters, and transported back to shore. Then, they must be treated with chemicals to extract the manganese. At the moment, this is all too costly and the manganese nodules on the ocean floors remain largely untouched.

Revolving crane

Flare

Well heads
Transport oil and gas to the platform.

Processing equipment

DRILLING FOR GOLD

An oil platform is like a huge steel and concrete hotel in the middle of the ocean. Several hundred workers can live there for weeks at a time. Most platforms stay in position for about 25 years, although one rig has survived for 60 years. Every day, these platforms pump millions of barrels of oil or "black gold," a name often given to this sticky, black, expensive liquid.

Lifeboat

CHECKING THE FOUNDATIONS

Divers must make routine checks of all the pipes and cables under an oil rig. Pounding seas constantly batter the rig and can eventually wear away or dislodge even the strongest metal framework.

DRILLING DEEP

Several oil wells extend at different angles through the layers of strata until they reach a pocket of oil.

Water level
The water level of the ocean rises and falls with the tides.

Shales and porous rocks
Water passes through the widely spaced rock grains of shale, sandstone and limestone.

Impermeable layer
A layer of dense rock forms a lid on top of the layer of oil or gas.

An oil and water mix
Oil, water and gas form pockets in the grains of porous rocks. Small amounts of oil and water mix together here.

Non-porous rocks
This layer, which is often granite, stops the oil from escaping downwards.

Q: How is oil formed?

• EXPLOITING THE OCEANS •

The Perils of Pollution

Human and industrial waste has been dumped in the sea for years. People once believed that sea water could kill any lurking germs. But scientists have now discovered that sewage pumped into the sea can spread terrible diseases, such as cholera, typhus and hepatitis. If polluting material enters the food chain, it can become more and more dangerous as one creature eats another. Humans are at the end of the food chain and they can suffer the worst effects of all. In the 1950s, many people in Japan died or became paralyzed after eating fish that was contaminated with mercury from a local factory. The United Nations began to take marine pollution seriously in the 1970s, but by this time, parts of some seas were already dying. Today, oceans are still polluted by oil spills, chemicals and human waste.

A GIANT GARBAGE CAN
Humans throw all kinds of litter into the sea, but this garbage never goes away.

A HAZARDOUS LIFE
This seal is entangled in a carelessly discarded fishing line. Any material thrown into the sea is potentially harmful to sea creatures. Nets, plastic balloons and the ring openers of soft-drink cans kill many sea mammals, sea birds and fish.

STRANGE BUT TRUE
This hermit crab was found with a very unnatural shell— a plastic bucket in which it is now trapped.

Spilling Oil

When an oil tanker spills its liquid cargo into the sea, floating booms are erected. They try to trap the oil and stop it from polluting the shoreline. The trapped oil is sucked up and stored somewhere safe. But it is not always possible to stop the oil from spreading. These workers are hosing the rocky coastline of Alaska in the huge clean-up operation that followed the *Exxon Valdez* oil spill.

BLUE-GREEN ALGAE
Clouds of marine algae, which grow when water is polluted, float near the surface of the Sea of Cortez. The algae shade the sea bottom and stop the rich seagrasses from growing. Without this source of food, fish, shellfish and worms will suffer.

THE BEGINNING
This undeveloped beach in Cyprus is already littered with trash from the ocean.

LOSING LIFE
Oil spills at sea are devastating for wildlife. When a bird's feathers are covered with oil, they are no longer waterproof. Water soaks into the unprotected feathers and the bird drowns or freezes to death.

• EXPLOITING THE OCEANS •

Conserving the Oceans

Each year there are 90 million more people on Earth to feed. Because the biggest increase in population is in coastal areas, more seafood is needed each year to feed the world. Sadly, most of the traditional fisheries of the oceans have either reached the limit of safe fishing or are already past that point and are now being overfished. If we want to keep the great ocean fisheries at their most productive, we must stop polluting them and also be more moderate in the numbers of fish we catch. This much-needed increase in the numbers of fish and shellfish can only happen in clean, unpolluted waters. Numbers can be further increased by opening new fisheries and by setting up fish farms in coastal lagoons and ponds, or within big nets in shallow water.

MARINE PARKS

Coral reefs are breaking down all over the world from the effects of pollution and overfishing. To prevent this, most tropical countries are creating marine parks to help protect coral areas from further damage. Tourists can still enjoy the reefs and their fishes at these parks. The largest of these sanctuaries is the Great Barrier Reef Marine Park. It extends for 1,240 miles (2,000 km) along the coast of northeastern Australia.

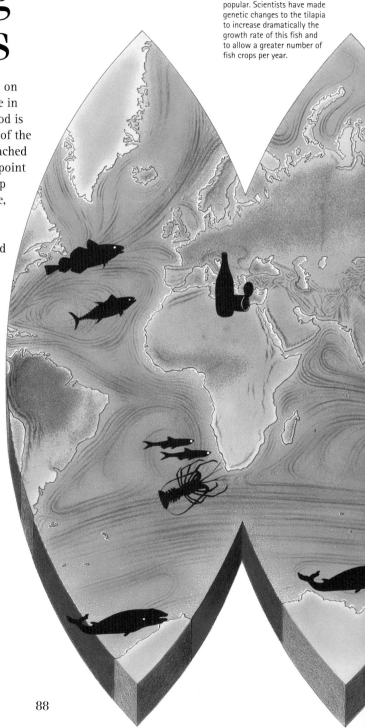

African Tilapia
Fish farming is a huge and growing industry in southeast Asia. Many local species are used, but African tilapias are very popular. Scientists have made genetic changes to the tilapia to increase dramatically the growth rate of this fish and to allow a greater number of fish crops per year.

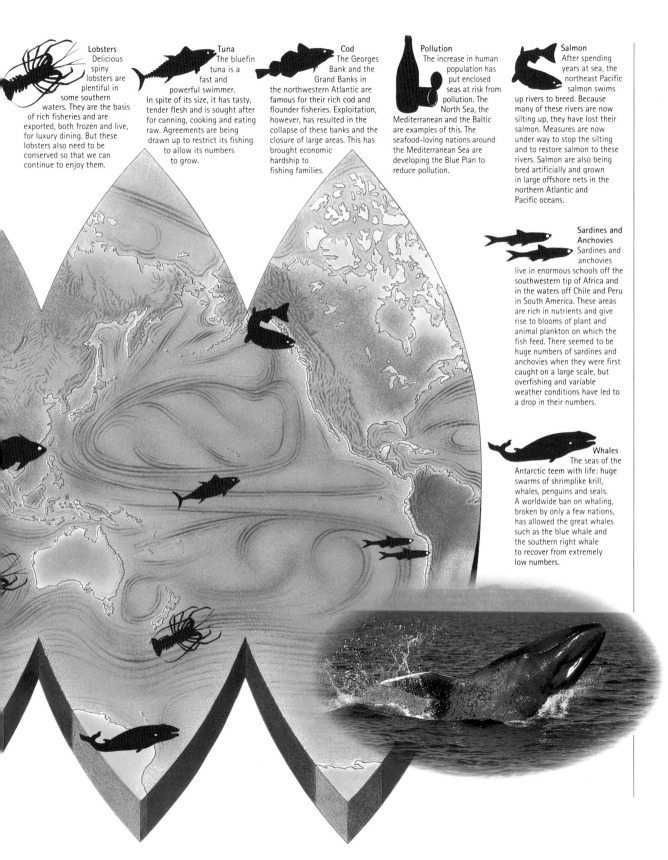

Lobsters Delicious spiny lobsters are plentiful in some southern waters. They are the basis of rich fisheries and are exported, both frozen and live, for luxury dining. But these lobsters also need to be conserved so that we can continue to enjoy them.

Tuna The bluefin tuna is a fast and powerful swimmer. In spite of its size, it has tasty, tender flesh and is sought after for canning, cooking and eating raw. Agreements are being drawn up to restrict its fishing to allow its numbers to grow.

Cod The Georges Bank and the Grand Banks in the northwestern Atlantic are famous for their rich cod and flounder fisheries. Exploitation, however, has resulted in the collapse of these banks and the closure of large areas. This has brought economic hardship to fishing families.

Pollution The increase in human population has put enclosed seas at risk from pollution. The North Sea, the Mediterranean and the Baltic are examples of this. The seafood-loving nations around the Mediterranean Sea are developing the Blue Plan to reduce pollution.

Salmon After spending years at sea, the northeast Pacific salmon swims up rivers to breed. Because many of these rivers are now silting up, they have lost their salmon. Measures are now under way to stop the silting and to restore salmon to these rivers. Salmon are also being bred artificially and grown in large offshore nets in the northern Atlantic and Pacific oceans.

Sardines and Anchovies Sardines and anchovies live in enormous schools off the southwestern tip of Africa and in the waters off Chile and Peru in South America. These areas are rich in nutrients and give rise to blooms of plant and animal plankton on which the fish feed. There seemed to be huge numbers of sardines and anchovies when they were first caught on a large scale, but overfishing and variable weather conditions have led to a drop in their numbers.

Whales The seas of the Antarctic teem with life: huge swarms of shrimplike krill, whales, penguins and seals. A worldwide ban on whaling, broken by only a few nations, has allowed the great whales such as the blue whale and the southern right whale to recover from extremely low numbers.

Volcanoes and Earthquakes

- What is a hotspot?
- Why does Iceland have many volcanoes?
- Where does lava come from?
- What is a tsunami?

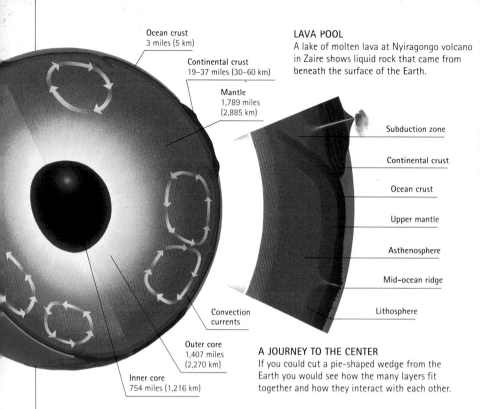

Ocean crust 3 miles (5 km)
Continental crust 19–37 miles (30–60 km)
Mantle 1,789 miles (2,885 km)
Convection currents
Outer core 1,407 miles (2,270 km)
Inner core 754 miles (1,216 km)

LAVA POOL A lake of molten lava at Nyiragongo volcano in Zaire shows liquid rock that came from beneath the surface of the Earth.

Subduction zone
Continental crust
Ocean crust
Upper mantle
Asthenosphere
Mid-ocean ridge
Lithosphere

A JOURNEY TO THE CENTER If you could cut a pie-shaped wedge from the Earth you would see how the many layers fit together and how they interact with each other.

• THE UNSTABLE EARTH •

Fire Down Below

The Earth is made up of several layers. If you could stand at the center, 3,950 miles (6,371 km) down, you would see the solid iron inner core, which is surrounded by an outer core of liquid iron and nickel. To travel to the surface, you would pass through the solid rock of the lower mantle, then the soft, squishy area of rock called the asthenosphere. The final two layers join together to form the lithosphere, which is made up of the solid rock of the upper mantle, and the crust. The crust covers the Earth as thin apple peel covers an apple. Inside the Earth, radioactive elements decay and produce heat, and the temperature increases to an amazing 5,432°F (3,000°C) at the core. This heat provides the energy for the layers to move and interact. Melted rock called magma rises from deep within the Earth to near the surface. Some of it cools and becomes solid within the crust but some erupts on the surface as lava.

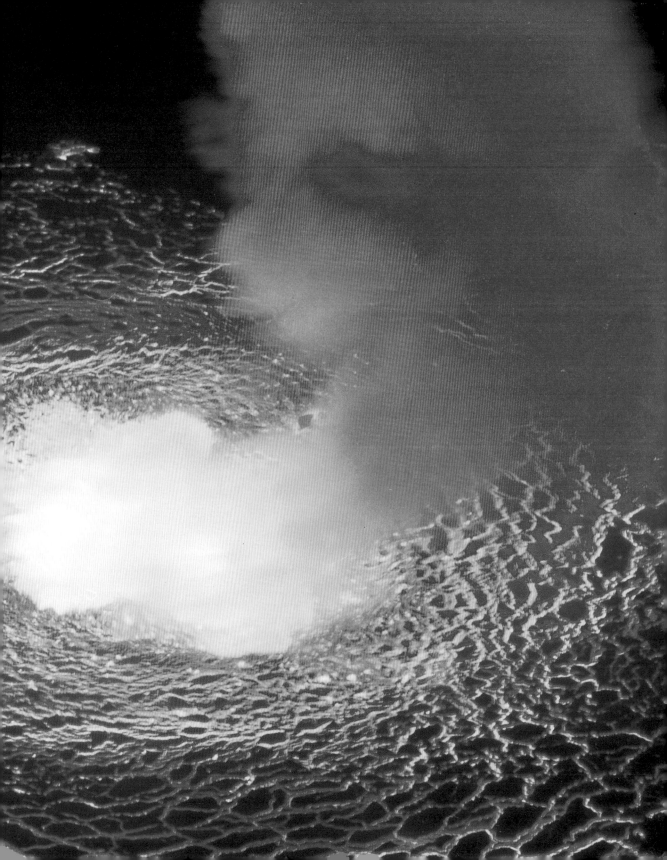

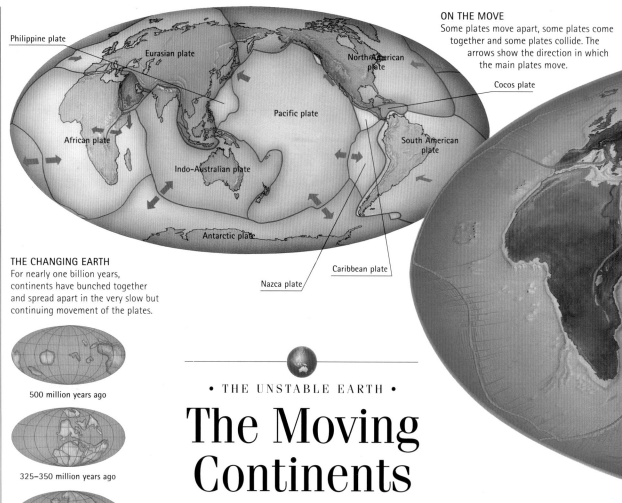

ON THE MOVE
Some plates move apart, some plates come together and some plates collide. The arrows show the direction in which the main plates move.

THE CHANGING EARTH
For nearly one billion years, continents have bunched together and spread apart in the very slow but continuing movement of the plates.

500 million years ago

325–350 million years ago

Pangaea the supercontinent
200 million years ago

Pangaea splits into
Gondwana and Laurasia
130 million years ago

Gondwana and Laurasia
65 million years ago

• THE UNSTABLE EARTH •

The Moving Continents

The Earth's outermost section, the lithosphere, is separated into seven large and several small jagged slabs called lithospheric plates, which fit together much like puzzle pieces. The crust, or top part, of each two-layered plate comes from either an ocean, a continent or a bit of both. You cannot feel it, but the plates are constantly moving. Supported by the soft, squishy material of the asthenosphere under them, plates pull and push against each other at a rate of $3/4$–8 in (2–20 cm) per year. When plates pull apart, magma from the mantle erupts and forms new ocean crust. When they move together, one plate slowly dives under the other and forms a deep ocean trench. Mountain ranges form when some plates collide. Other plates slide and scrape past each other. Most volcanoes and earthquakes occur along edges where plates meet.

A NEW THEORY
In 1915, Alfred Wegener, a German scientist and explorer, proposed his theory of continental drift. He wrote of a huge supercontinent, Pangaea, which split apart millions of years ago. The pieces then slowly drifted to their present position. Not until the mid-1960s was the theory revised and accepted by scientists.

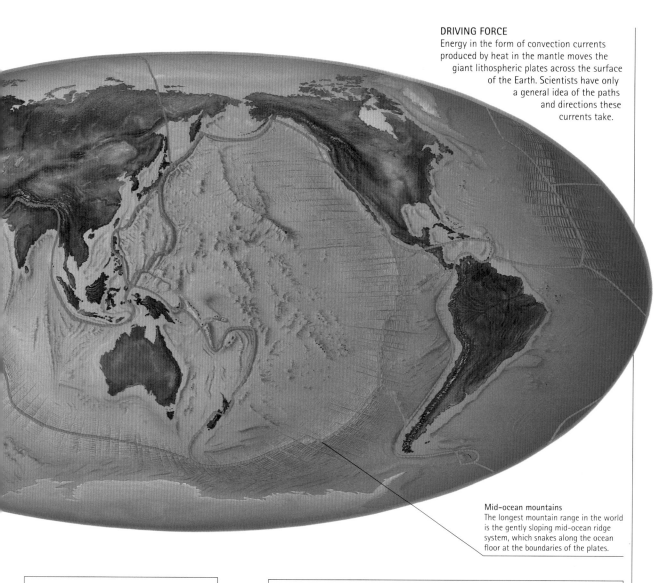

DRIVING FORCE
Energy in the form of convection currents produced by heat in the mantle moves the giant lithospheric plates across the surface of the Earth. Scientists have only a general idea of the paths and directions these currents take.

Mid-ocean mountains
The longest mountain range in the world is the gently sloping mid-ocean ridge system, which snakes along the ocean floor at the boundaries of the plates.

STRANGE BUT TRUE
Many millions of years ago, the east coast of South America and the west coast of Africa fitted together as part of the same land mass. Today, the two continents even have similar rock, plant and animal fossils.

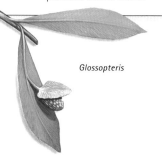

Glossopteris

SLIDING BY

When some plates meet they slide past each other in opposite directions or in the same direction at different speeds. The edges of the sliding plates grind when they meet, causing a series of weblike cracks or faults and fractures to occur. Here an orchard changes course to follow the line of a fault—a simple example of plate movement. As the two plates move slowly past each other, stress builds up in the rocks below and an earthquake can occur.

Discover more in Fire Down Below

• THE UNSTABLE EARTH •

Ridges and Rift Valleys

The ocean crust is rugged with mountain ridges and deep rift valleys or cracks. When two plates with ocean crust move apart, magma from the mantle bubbles up to the surface to fill the rift. The magma cools and hardens and adds new strips of crust or ocean floor to the edges of the two plates. This forms what is called a spreading, or widening, ridge. The Atlantic Ocean is widening by 3/4 in (2 cm) per year. The East Pacific Rise is widening by 8 in (20 cm) per year—the fastest widening rate of an ocean floor. In 10 million years, it will be 1,240 miles (2,000 km) wider. Plates continue to move away from the spreading ridge towards other plates. When a spreading ridge fractures or breaks, earthquakes occur. Mid-ocean ridge volcanoes also form in rifts, fed by the magma from below. Over millions of years, they can grow so large that they rise above the water to form islands, such as Iceland on the North Atlantic mid-ocean ridge.

Smoking chimneys
Sulfur and other minerals deposited on the sides of the vents build natural chimneys of up to 33 feet (10 m) high.

Q: How do oceans grow wider?

SEA FLOOR SPREADING
New lithospheric plate is created at a spreading ridge.

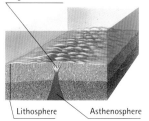

As two plates with ocean crust move apart, a crack or rift forms.

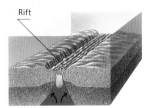

Magma from the mantle rises to fill the rift between the two plates.

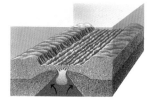

Magma cools and hardens, and adds to the edges of the plates.

BLACK SMOKERS
Volcanic hot springs or hydrothermal vents such as these were discovered in 1977 and are found along very active spreading ridges on the ocean floor. The mineral-rich water is eaten by bacteria, which is food for other vent animals such as tube worms, giant clams and eelpouts.

Circulation system
The arrows show how cold sea water seeps down through the ridge. Magma heats it to an amazing 572°F (300°C), then the water rises back up through the vent.

An Island is Born

In 1963, an undersea volcanic eruption created Surtsey Island, the newest land mass on Earth. Surtsey lies off the southwest coast of Iceland, a country known for its many active volcanoes. The eruption began with a large column of ash and smoke. Heat and pressure from deep within the Earth pushed part of the mid-Atlantic ridge to the surface. The island continued to grow for several months. Today, Surtsey Island measures 1 sq mile (2.6 sq km). Its dominant feature is the steep cone of the volcano.

Discover more in The Moving Continents

• THE UNSTABLE EARTH •

Subduction

When two plates move toward each other and meet, one plate slowly dives, or subducts, beneath the other along what is called a "subduction zone." This is an area of intense earthquake and volcanic activity caused by the movement of the two plates. There are three types of plate boundaries where subduction occurs: ocean to ocean, ocean to continent and continent to continent. The action of subduction is the same for all three types but the results are different. The area at which one plate dives beneath another creates an ocean trench, the deepest part of the ocean floor. As the down-going plate continues to sink deeper into the mantle, it mixes with hot rock and melts to form magma. Under extreme heat and pressure, this new magma mixture then forces its way back upwards to erupt violently at the surface. Compared to the gentle chain of mid-ocean ridge volcanoes, subduction volcanoes are explosive and dangerous. This is due to the presence of water and the build-up of gases dissolved in the thick and sticky magma that is produced in the process of subduction.

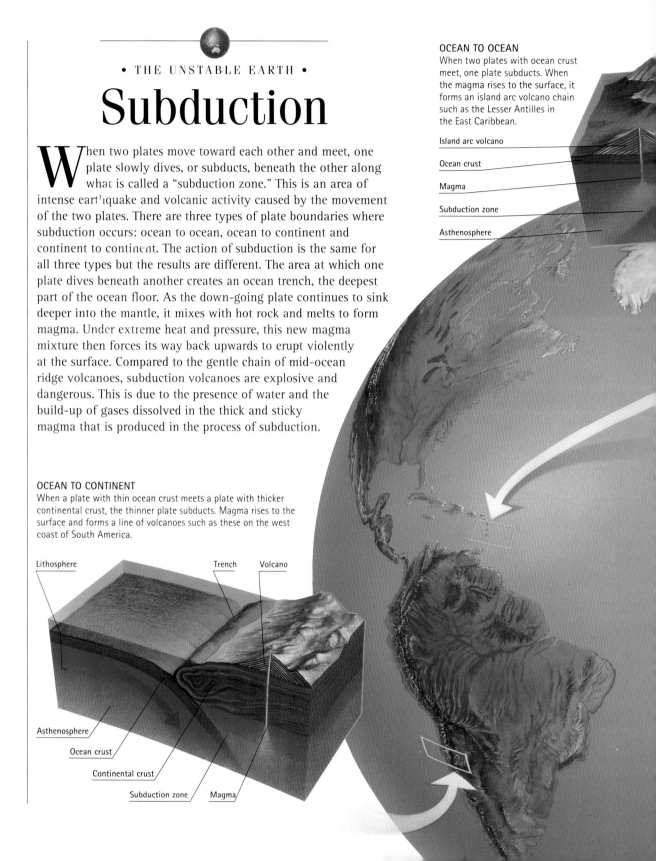

OCEAN TO OCEAN
When two plates with ocean crust meet, one plate subducts. When the magma rises to the surface, it forms an island arc volcano chain such as the Lesser Antilles in the East Caribbean.

- Island arc volcano
- Ocean crust
- Magma
- Subduction zone
- Asthenosphere

OCEAN TO CONTINENT
When a plate with thin ocean crust meets a plate with thicker continental crust, the thinner plate subducts. Magma rises to the surface and forms a line of volcanoes such as these on the west coast of South America.

- Lithosphere
- Trench
- Volcano
- Asthenosphere
- Ocean crust
- Continental crust
- Subduction zone
- Magma

Trench Ocean crust Lithosphere

Q: What can happen when two plates meet?

RING OF FIRE

The Pacific Ring of Fire is the name scientists use to describe the area of the world located around the edges of the Pacific Ocean. This is actually the boundary of the large Pacific plate, which is slowly moving and subducting under or grinding past other plates it meets. In 1994, two volcanoes, Mt. Tavurvur and Mt. Vulcan, erupted in Rabaul, Papua New Guinea. These island arc volcanoes formed at an area where the Pacific plate is subducting under the South Bismarck ocean plate.

WHERE IN THE WORLD?
This map shows the general locations in the world where one plate is subducting under another.

CONTINENT TO CONTINENT
When these plates move towards each other, the thin ocean part of one plate subducts but the continental part continues to push against the other plate and forms a mountain chain such as the Himalayan Mountains in central Asia.

Mountains Subduction zone Continental crust Lithosphere

Asthenosphere

Ocean crust

Discover more in The Moving Continents

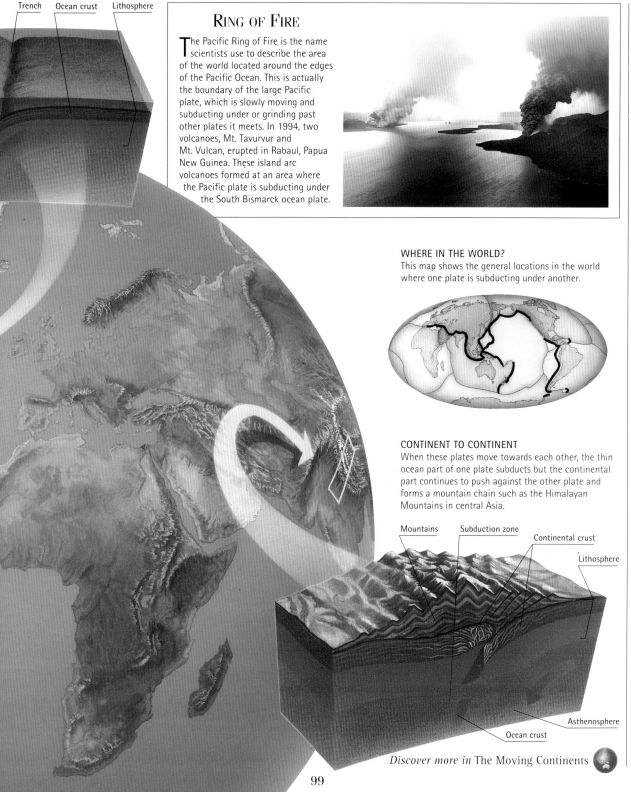

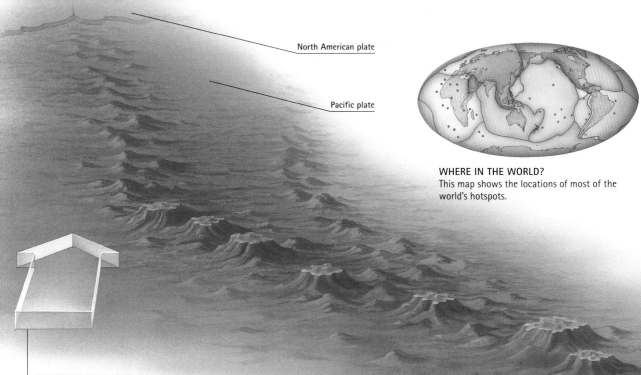

North American plate

Pacific plate

WHERE IN THE WORLD?
This map shows the locations of most of the world's hotspots.

DIRECTION OF PLATES
The Pacific plate, moving northwest, has slowly carried the oldest of the Hawaiian Islands away from the hotspot and closer to the point where they will subduct under the North American plate.

• THE UNSTABLE EARTH •

Hotspots

Volcanoes are born in different ways and hotspot volcanoes, though spectacular, are often less violent than those that occur at subduction zones. Hotspot volcanoes form in the middle of plates, directly above a source of magma. Molten rock rises to the surface from deep within the Earth's mantle, pierces the plate like a blow torch and erupts in a lava flow or fountain. A hotspot stays still, but the plate keeps moving. Over millions of years this process forms a string of volcanic islands such as the Hawaiian Island chain on the Pacific plate. Hawaii's Mauna Loa and Kilauea, now active volcanoes, will gradually become dormant or cold as Hawaii moves off the hotspot. A new active volcano, such as Loihi to the southeast, might become a new island above the hotspot. The Pacific plate has carried other island volcanoes in the chain far away from the magma source.

INDIAN OCEAN VOLCANO
Piton de la Fournaise on Réunion Island is one of the most active volcanoes in the world.

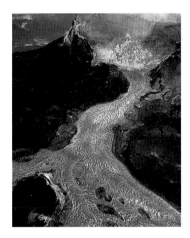

KILAUEA FLOWS
The lava that erupts from hotspot volcanoes such as Kilauea in Hawaii can move at speeds of up to 62 miles (100 km) per hour.

Hotspot, Continental Style

Yellowstone National Park, in Wyoming, has spectacular geysers and natural hot springs and is one of the most famous hotspots found on a continent. Underground water is heated by a hotspot source of magma deep within the mantle. The steam in the boiling water expands, and water and steam burst through the many cracks in the crust and erupt as geysers. Some scientists believe that one day, in the next few hundred thousand years, a major volcanic eruption could occur in the area.

Q: What happens to a hotspot volcano as it moves away from a source of magma?

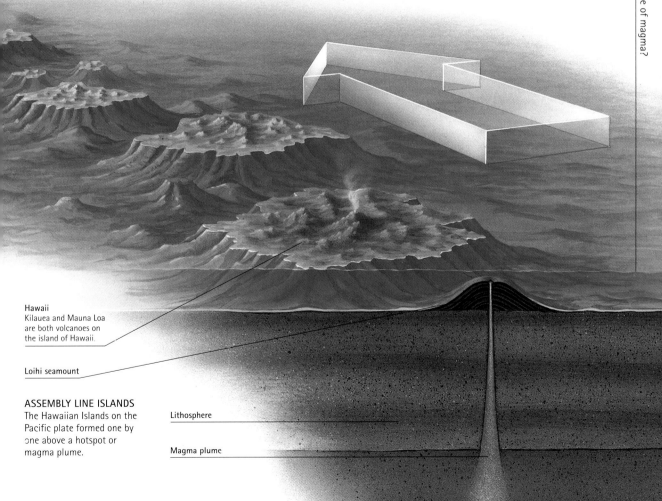

Hawaii
Kilauea and Mauna Loa are both volcanoes on the island of Hawaii.

Loihi seamount

ASSEMBLY LINE ISLANDS
The Hawaiian Islands on the Pacific plate formed one by one above a hotspot or magma plume.

Lithosphere

Magma plume

TYPES OF VOLCANIC ERUPTION

HAWAIIAN
Large amounts of runny lava erupt and produce large, low volcanoes.

PELÉEAN
Blocks of thick, sticky lava are followed by a burning cloud of ash and gas.

STROMBOLIAN
Small, sticky lava bombs and blocks, ash, gas and glowing cinders erupt.

VULCANIAN
Violent explosions shoot out very thick lava and large lava bombs.

PLINIAN
Cinders, gas and ash erupt great distances into the air.

• VOLCANOES •

Volcanic Eruptions

Deep inside the Earth, magma rises upwards, gathers in pools within or below the crust and tries to get to the surface. Cracks provide escape routes, and the magma erupts as a volcano. Steam and gas form clouds of white smoke, small fragments of rock and lava blow out as volcanic ash and cinder, and small hot bombs of lava shoot out and harden. Not all lava is the same. It may be thick and sticky or thin and runny. Lava thickness, or viscosity, determines the type of volcanic eruption and the kind of rock that forms when the lava hardens. Some volcanoes are active, erupting at any time; some are dormant or cold, waiting to erupt; others are dead or extinct. Volcanoes have shaped many of the Earth's islands, mountains and plains. They have also been responsible for changing weather, burying cities and killing people who live nearby.

THE INSIDE STORY
This cross-section shows the inner workings of a volcano and what happens during an eruption.

Side vent
Under pressure, this side vent branches off from the central vent and carries lava upwards through cracks in the rocks to ooze out the side of the volcano.

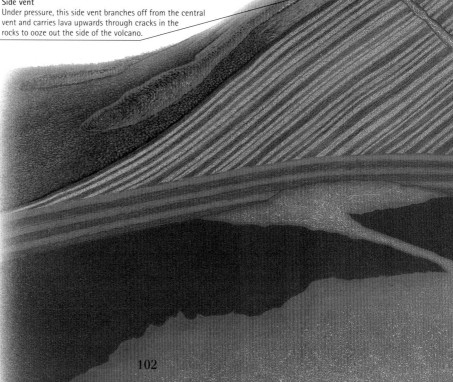

Why People Live Near Volcanoes

For centuries communities have grown up in the shadows of volcanoes. In Iceland, people use the energy from their island's many volcanoes to provide heat and power. Other people live near volcanoes because the soil is rich, and farmers grow crops and graze their herds on the slopes. In Indonesia more people live on the islands with active volcanoes than on the islands with none. Shown here are lush rice terraces growing on the fertile ground near the volcano Mt. Agung in Bali, Indonesia.

Q: What can happen to magma when it rises to the surface?

From the top
A white, smoke-like mixture of steam, ash and gas is blown into the air. Hard bits of lava called bombs shoot out from the top, while molten lava flows down the sides of the volcano.

Crater
This funnel-shaped opening at the top of the volcano enables lava, ash, gas and steam to erupt.

Cone
The cup-shaped cone is built up by ash and lava from a number of eruptions.

Central vent
The main vent, or chimney, rises from the magma chamber below. Magma flows up the vent to erupt on the surface as lava.

Sill
Magma does not always find an outlet to the surface. Some gathers, cools and becomes solid between two underground layers of rock.

Magma chamber
Thick, molten magma travels upward from the mantle and collects in large pockets in the crust where it mixes with gases and water. Under pressure from heat in the mantle the magma forces its way through vents to the surface.

Fissure eruption
Some magma forces its way upward through vertical cracks in the rock and erupts on the surface.

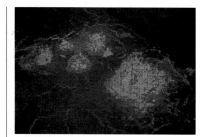

LAVA LAKES
Lava collects in a series of pools after the eruption of Pu'u O'o in Hawaii.

INTO THE SEA
Runny lava gushes from the side of a Hawaiian volcano into the sea below, where it cools.

• VOLCANOES •

Lava Flows

Lava is red-hot magma that erupts at the Earth's surface. A river of runny lava flows downhill from a volcano's central crater or oozes from a crack in the ground. It cools down from temperatures of up to 2,192°F (1,200°C), becomes stickier and slower-moving and often forms volcanic rock called basalt. Explosive, killer subduction volcanoes erupt thick, sticky lava that contains large amounts of silica. This cools to form rocks such as rhyolite, and different forms of volcanic glass such as obsidian. Blocked outlets or large amounts of gas or water in the lava cause violent eruptions, with lava bombs and boulders shooting out in all directions. If lava hardens with a rough, broken surface, it is called "aa". Lava that hardens with a smoother covering skin is called "pahoehoe". Pahoehoe can wrinkle to form ropes of rock. Hardened lava often cracks to form regular columns.

RIVER OF FIRE
In this Hawaiian-type eruption, the lava streams down from the volcano into an area where it burns houses and adds a new layer of volcanic rock to the ground.

TYPES OF LAVA FLOW

Lava is named according to how it looks as it cools and hardens. Pillow lava is the most common form of lava found on Earth. It erupts in water especially along the mid-ocean ridge where volcanoes gently ooze pillow-shaped lumps of lava through cracks in the ocean floor. Pillow lava can also be found on dry land that was once part of the ocean floor. Pahoehoe lava is runny and very fast-flowing. The surface cools quickly and forms a thin, smooth skin with hot lava still moving underneath, which can twist and coil the surface to look like rope. Aa lava flows move more slowly and are not as hot as pahoehoe. Aa flows cool to form sharp chunks of rock, which can be as thick as 328 feet (100 m).

Pillow lava

Pahoehoe lava

Aa lava

ASH CLOUD
Ash shoots high in the sky in explosive subduction volcanoes. In a pyroclastic flow, the ash cloud drops and speeds down the side.

SPREADING ASH
Present-day Santorini in Greece (indicated by the red dot) was the site of an eruption in 1645 BC. Scientists believe that a 18-mile (30-km) high column of ash spread over a large part of the east Mediterranean area (as shown).

• VOLCANOES •

Gas and Ash

As magma rises to the Earth's surface, the gases mixed in with it expand or swell and try to escape. These gases sometimes contain carbon dioxide and hydrogen sulphide, which are both very harmful to humans. In runny magma, gases have no trouble escaping and cause mild eruptions with a lava flow. Gases trapped in thick, sticky lava build up and explode violently. These explosive eruptions fling clouds of rock fragments and hardened lava froth called pumice several miles into the air. Ash is formed during the explosion when rock and lava blow apart into millions of tiny pieces. Falling ash spreads further and does more damage than lava flows do, although a small sprinkling of ash can add important nutrients to the soil. Sometimes the wind carries clouds of ash around the world, affecting weather patterns.

COVERED IN ASH
Residents in protective masks walk down a street in Olongapo City in the Philippines after the ash-fall from the 1991 eruption of Mt. Pinatubo.

Maurice and Katia Krafft

French volcanologist Maurice Krafft and his geochemist wife Katia were responsible for some of the world's most spectacular volcano photographs. They witnessed more than 150 volcanic eruptions throughout the world and wrote many books and films. Their field work often placed them in great danger, and they had to wear heat-resistant suits for close-up photography. Both died in 1991 while they were filming a pyroclastic flow from the eruption of Mt. Unzen in Japan.

PYROCLASTIC ROCKS
Small fragments of volcanic rock and frothy pumice fly into the air. Larger bombs or blocks, some the size of boulders, bounce down the side of the mountain.

Q: What makes ash from a volcano dangerous?

PYROCLASTIC FLOW
This rapidly moving avalanche of volcanic fragments and gases, at a temperature of 212° F (100° C), can rocket down the side of a volcano at speeds of up to 155 miles (250 km) per hour, destroying everything in its path.

WHEN A CALDERA FORMS
During mild explosions, magma rises to the top of the volcano's main vent.

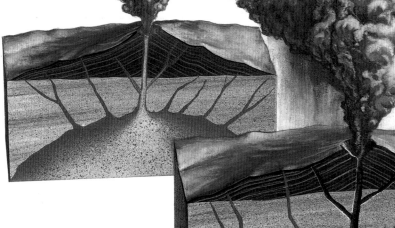

THE NEXT STAGE
As the eruption increases in strength, the magma rapidly sinks back down to the top of the magma chamber.

ERUPTING CALDERA
A cloud of ash and gas erupts from the caldera of Mt. Ngauruhoe in New Zealand.

THE CLIMAX
In Plinian or Peléean eruptions, the activity climaxes and magma sinks below the roof of the magma chamber, leaving an empty space where it once supported the roof.

• AFTER THE EVENT •

Craters and Calderas

Craters are the funnel-shaped hollows or cavities that form at the openings or vents of volcanoes. The simplest craters occur on the top of cones and usually have a diameter of about 6/10 mile (1 km) or less. Volcanoes can also form craters at the side. Small lava lakes occur when the lava is unable to escape, and it blocks the vent like a bath plug. Calderas are very large craters formed by an explosion or massive volcanic eruption. The magma chamber empties and can no longer support the weight of the volcano and the cone collapses. Calderas are often more than 3 miles (5 km) in diameter. The world's largest caldera is at Aso, Japan, and it is 14 miles (23 km) long and 10 miles (16 km) wide. When a volcano is dormant or extinct, a caldera can fill with water to form a large lake.

KILAUEA CRATER
Lava erupts from the crater of Kilauea volcano on Hawaii.

CRATER LAKES

These lakes form when the main vents of dormant or extinct volcanoes are plugged with hardened lava or other rubble. Over many years the crater gradually fills with water from rain or snow. Shown here is Crater Lake in Oregon. This lake is a caldera that formed when the summit of Mt. Mazama collapsed more than 6,600 years ago. The small cone within the caldera is called Wizard Island.

LIFE IN A CALDERA
Pinggan Village is one of the many villages within the caldera of the extinct volcano Gunung Batur, in Bali, Indonesia.

THE COLLAPSE
Once the magma support is removed, the top collapses into the magma chamber, and more eruptions can occur on the caldera floor.

STRANGE BUT TRUE

On the Indonesian island of Flores, the volcano of Keli Mutu is well known for its different colored crater lakes. Tiwoe Noea Moeri Kooh Pai is green, Tiwoe Ata is light green and Tiwoe Ata Polo is an amazing red color.

Q: What can make the cone of a volcano collapse?

Discover more in Volcanic Eruptions

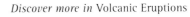

VOLCANIC PLUG
This plug in Cameroon, West Africa, started as magma in the vent of a volcano. Over millions of years the magma cooled and hardened. The softer rocks eroded and left the plug exposed.

• AFTER THE EVENT •

Volcanic Rocks and Landforms

Deep within the Earth the heat is so intense that the red-hot rock or magma is molten. This magma rises toward the surface through cracks in the crust, then cools and hardens to form igneous rocks. When these rocks cool slowly underground within the cracks, they form intrusive igneous rocks such as granite. Intrusive rocks may appear on the surface when the surrounding softer rocks erode and expose amazing landscapes with landforms such as sills, dykes and plugs. Magma that erupts at the surface as lava and cools quickly becomes extrusive igneous rock. Runny lavas produce a rock called basalt, which in large quantities can flood an area and make basalt plateaus. Thicker, stickier lava can produce pumice, volcanic glass such as obsidian, and a light-colored rock called rhyolite.

ROCK WALL
Most volcanoes are fed by magma that forces its way up through vertical cracks in the rocks. The magma can harden underground to form a wall of rock called a dike. Erosion has exposed this dike in eastern Australia, which is called the "Breadknife."

HOT SPRINGS
Hot springs form when water seeps down through rocks and is heated by magma or hot rock from below. The water then rises up towards the surface. This Japanese macaque enjoys a warm soak in one of the hot springs in northern Japan—a volcanic area near the boundaries of the Pacific and Philippine plates.

Did You Know?

Diamonds originate in the mantle under extreme heat and pressure. Diamonds, such as this one embedded in volcanic rock called kimberlite, are pushed to the surface by rising magma.

Useful Volcanoes

There are many uses for volcanic rocks. Small amounts of ash can add nutrients to soil. Basalt and granite are useful building materials. The mineral sulfur grows around the vents of some active volcanoes. Sulfur is mined for use in manufacturing. It is added to rubber to make it last longer, and is also an ingredient in many explosives. Here workers carry baskets full of large sulfur crystals collected from around the vent of Kawah Ijen volcano in Java, Indonesia. This sulfur will be processed and mixed with phosphate to make a type of fertilizer.

Q: What kinds of rock come from a volcano?

CAPPADOCCIA
This unusual volcanic landscape formed at Cappadoccia, Turkey, from lava eruptions of the now extinct Mt. Erciyes. Wind and rain have eroded the volcanic rock into pointed formations.

ETNA ERUPTS
A column of ash, gas and lava shoots into the air during a 1992 eruption of Mt. Etna.

THE SLOPES OF MT. ETNA
An old illustration shows the lava flow from one of the four main volcanic vents.

• FAMOUS VOLCANOES •

Mediterranean Eruptions

For more than two million years, earthquakes and volcanic eruptions have occurred in the area of the Mediterranean Sea along the boundary where the African plate meets the Eurasian plate. The Bay of Naples, in Italy, is the site of Mt. Vesuvius, which erupted violently in AD 79 and destroyed the towns of Herculaneum and Pompeii. Pliny the Younger wrote the very first account of an eruption after observing the event. Since then, Mt. Vesuvius has erupted numerous times. The last eruption was in 1944 but it is not known how long the volcano will remain dormant. Mt. Etna dominates eastern Sicily and is Europe's largest active volcano. It has been erupting periodically for more than 2,500 years and regularly destroys villages and farmland. People continue to settle on its fertile soil. The last major eruption of Mt. Etna occurred in 1992.

ON THEIR DOORSTEP
Mt. Vesuvius looms above the modern city of Naples. When will it erupt next?

DID YOU KNOW?
The 1944 eruption of Mt. Vesuvius occurred during the Second World War. Glass-sharp volcanic ash and rock fragments seriously damaged aircraft engines.

RAINING ASH
People fled the town of Herculaneum during the AD 79 eruption of Mt. Vesuvius. Some ran towards the sea and escaped in boats, but many perished when a hot surge of ash and gas covered the town.

A Greek Tragedy

The beautiful ancient Greek island of Thera (today called Santorini) in the Aegean Sea was destroyed by a violent volcanic eruption 3,500 years ago. The island was home to the Minoan people, a very wealthy and advanced civilization. The eruption caused huge tsunamis and ash-falls. These swept across neighboring islands such as Crete, site of the Minoan capital of Knossos, where this vase was found. What was left of Thera was covered in more than 197 ft (60 m) of ash and pumice.

DID YOU KNOW?
Although the 1815 eruption of Tambora in Indonesia was a much larger and more violent eruption, Krakatau seized the entire world's attention because of modern communication networks.

Over England
The evening sky over London dazzled onlookers with beautiful colors. Waves raised the tides in the English Channel.

Weather changes
Volcanic dust circled the globe for several years and lowered the Earth's average temperature. Hawaiians noticed a white halo around the sun.

Trinidad
On the other side of the globe, in Trinidad, the sun appeared blue.

• FAMOUS VOLCANOES •

Krakatau

In May, 1883, ash, gas and pumice erupted from a volcano on the Indonesian island of Krakatau. The island is located in an unstable area where the Indo-Australian plate subducts under the Eurasian plate. The early rumblings from the volcano were just warm-ups before the violent explosion that blew the island apart on August 27. The boom from the explosion, one of the loudest ever recorded, was heard 2,170 miles (3,500 km) away. Clouds of dust and ash rose 50 miles (80 km) into the air, circled the globe, and created many colorful sunsets around the world. As the volcano collapsed in on itself, giant waves called tsunamis rose more than 131 ft (40 m) high. These walls of water surged into 163 villages along the coastlines of Java and Sumatra, destroying them and killing almost 36,000 villagers. Floating islands of pumice endangered ships sailing in the Indian Ocean.

NEARING THE END
This nineteenth-century engraving was based on an old photograph taken in 1883, just three months before Krakatau exploded.

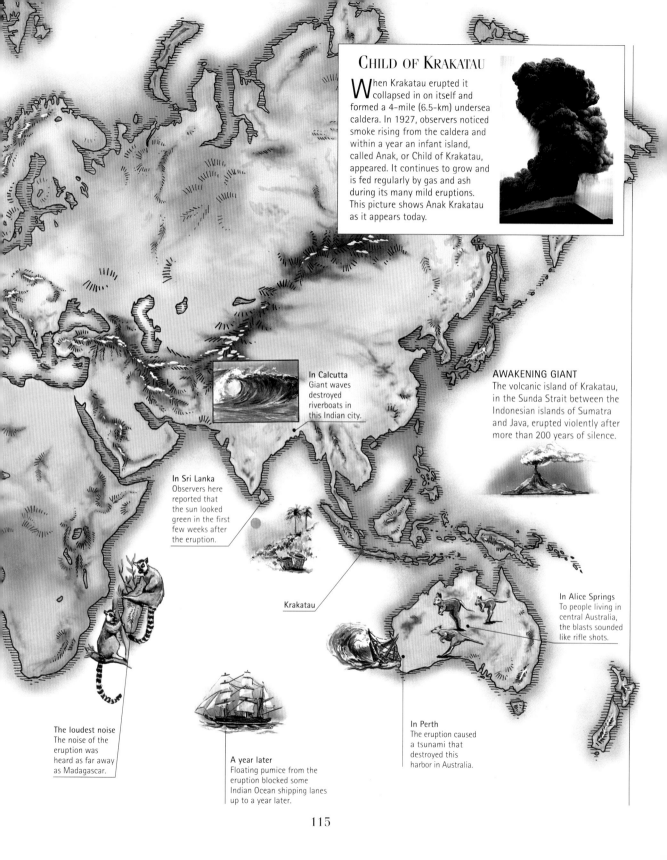

CHILD OF KRAKATAU

When Krakatau erupted it collapsed in on itself and formed a 4-mile (6.5-km) undersea caldera. In 1927, observers noticed smoke rising from the caldera and within a year an infant island, called Anak, or Child of Krakatau, appeared. It continues to grow and is fed regularly by gas and ash during its many mild eruptions. This picture shows Anak Krakatau as it appears today.

In Calcutta
Giant waves destroyed riverboats in this Indian city.

AWAKENING GIANT
The volcanic island of Krakatau, in the Sunda Strait between the Indonesian islands of Sumatra and Java, erupted violently after more than 200 years of silence.

In Sri Lanka
Observers here reported that the sun looked green in the first few weeks after the eruption.

Krakatau

In Alice Springs
To people living in central Australia, the blasts sounded like rifle shots.

The loudest noise
The noise of the eruption was heard as far away as Madagascar.

A year later
Floating pumice from the eruption blocked some Indian Ocean shipping lanes up to a year later.

In Perth
The eruption caused a tsunami that destroyed this harbor in Australia.

• FAMOUS VOLCANOES •

Iceland

Iceland is an island country that sits astride the northern Atlantic section of the mid-ocean ridge. The island provides scientists with an ideal place to study the ocean ridge above water. One part of Iceland is on the North American plate, which is moving westward, and the other part is on the Eurasian plate, which is moving eastward. As the island is slowly pulled in two, a rift, or large crack, is forming. Ravines and cliffs mark the edges of the two plates. Magma rising to the surface has created a series of central volcanoes separated by groups of fissures. As the area becomes more unstable, there is more earthquake and volcanic activity. Icelanders use the geothermal energy from their volcanoes for central heating, hot water and other electrical power.

DRAMATIC DISPLAY
Eldfell volcano gives a dramatic light show behind this church in the seaport town of Vestmannaeyjar.

AN ISLAND ERUPTION
In January, 1973, the seaport town of Vestmannaeyjar on Heimaey Island became the site of a new volcano called Eldfell. Most residents were evacuated, but for six months volunteers stayed behind to save what they could of the town.

Lava flow
Residents armed with fire hoses sprayed water on the advancing lava that threatened to take over the harbor. They saved the harbor but not before the lava flow added another 1 sq mile (2.6 sq km) of new land to the island.

DID YOU KNOW?

Iceland is the source of one-fifth of the Earth's recorded lava output. Scientists believe this is partly due to Iceland's position over an active spreading ridge and, possibly, a hotspot.

LANDSCAPE OF FIRE

In 1783, the 3-mile (5-km) long Lakagígar fissure in southern Iceland began erupting huge fountains of lava and large amounts of gas and ash. The lava flow, one of the largest ever recorded on Earth, eventually covered more than 220 sq miles (565 sq km) of land. A deadly blue haze settled over the country and spread to parts of Europe and Asia. Nobody was killed by the lava itself, but Iceland's crops were destroyed and much of the livestock starved to death. More than 10,000 people died in the famine that followed. Since then, fissures such as the one at Krafla in northeastern Iceland have continued to erupt.

The craters of Lakagígar fissure

Krafla fissure

Q: Why does Iceland have so many volcanoes?

BACKYARD VOLCANO
The crater of Eldfell glowed red as thick lava flowed down the side of the volcano and fiery ash rained down on the abandoned houses.

BURIED IN ASH
Much of the town lay beneath a thick layer of ash. Here, volunteers clear ash from rooftops to prevent the houses from collapsing.

• FAMOUS VOLCANOES •

Mt. St. Helens

The volcano Mt. St. Helens is one of 15 in the Cascade Range of the northwest United States—an area where the Juan de Fuca plate is subducting beneath the North American plate. On March 20, 1980, a string of earthquakes northwest of the mountain peak signalled the slow awakening of the volcano, which had been dormant since 1857. A week later, a small eruption shot ash and steam into the air. Groups of scientists arrived with instruments to monitor the volcano. By early May, a bulge developed on the cone. This indicated magma rising in the volcano's vent. The bulge grew bigger each day until a violent explosion, probably triggered by another earthquake, blew out the northern side of the mountain on May 18. This caused an enormous landslide that devastated an area of 234 sq miles (600 sq km) and triggered mudflows and floods.

BEFORE
In early 1980, the beautiful snow-capped peak of Mt. St. Helens was surrounded by forests and lakes.

AND AFTER
The north side of the nearly perfect cone blew apart in minutes, causing one of the largest volcanic landslides ever recorded.

PLINIAN PLUME
Minutes after the first explosion a second eruption produced a large Plinian column of ash and gas that rose to a height of 12 miles (20 km). This phase of the volcanic eruption continued for nine hours.

LIKE MATCHSTICKS
More than six million trees were uprooted or flattened by rock blasted from the volcano. After a massive salvage operation to clear the logs, seedlings were planted to replace the forests.

MUDFLOWS
Thick, sticky mud caused by melting snow and ice sped down the North Toutle River Valley into communities below.

Sweeping Ash

Millions of tons of ash shot 15 miles (25 km) into the atmosphere and the falling ash spread more than 930 miles (1,500 km) to the east. Ash fell like black snow in parts of Montana, Idaho, Oregon and Washington and covered streets, cars and buildings. Cats and dogs downwind from the eruption turned pale grey from the ash that floated from the sky. But ash does not melt like snow, and it had to be cleared. Most ended up in landfills.

Q: What other volcanic eruptions produced effects similar to those caused by Mt. St. Helens?

• EARTHQUAKES •

Surviving an Earthquake

When an earthquake strikes, beware of falling buildings and flying objects. To prevent building collapse and loss of life, engineers in many earthquake-prone cities follow strict rules when they repair earthquake damage or put up new structures. Many buildings are now designed to rest on reinforced concrete rafts that float when shock waves pass through. Existing buildings can be reinforced by cross-bracing walls, floors, roofs and foundations to help them withstand forces striking them from all directions. Walls and ceilings are strengthened with plywood in case of fire. Some areas have flexible gas lines that bend but do not break under pressure. Heavy furniture is bolted to walls to prevent it from flying around a room. Warning systems monitor stress in the Earth's crust, but earthquakes are still unpredictable and often much stronger than expected. In some areas, earthquake-proof buildings have collapsed.

BE PREPARED
Earthquake drills are part of the daily life of school children in Parkfield, California. Part of the San Andreas fault system lies beneath this small town.

SAFETY MEASURES
In this school, computers are bolted to tables, bookshelves and cabinets are attached to walls or fastened together. Windows are covered in transparent tape to stop them from breaking during a tremor.

WHEN THE EARTH SHAKES
An earthquake alarm is attached to the school's propane gas supply. The alarm switches off the gas automatically if a tremor above 3.5 on the Richter scale rocks the building.

TRANSAMERICA BUILDING
This rocket-shaped building in San Francisco, California, is designed to withstand an earthquake. The base is built on a concrete raft that gives extra support.

DROP!
When the teacher shouts "drop," each child crouches under the nearest desk. They link one arm around the leg of the desk to anchor it, and cover their head with both hands.

Liquid Soil

Buildings constructed on loose wet soil, such as lake beds or filled land, are at a greater risk during an earthquake due to a process called liquefaction. When the ground shakes, the solid soil particles separate from each other and the soil itself becomes a thick, muddy liquid. Buildings sink or fall over because they have no support. Cities such as San Francisco and Tokyo, Japan, have building rules that require stronger foundations for new structures. The buildings pictured here toppled over during the 1985 earthquake in Mexico City, Mexico, as a result of liquefaction.

Q: What would you do if an earthquake occurred?

TSUNAMI DAMAGE
Indonesian fishermen try to save what they can from their house, destroyed by one of the 12 tsunamis that struck East Java in June 1994.

DID YOU KNOW?
In 1896, a tsunami hit the northeastern coast of Japan. Local fishermen returned home to a devastated harbour. They had not noticed the tsunami when it passed under their fishing boats far out at sea.

• EARTHQUAKES •

Tsunamis and Floods

Tsunamis, huge killer waves, are caused by a jolt to the ocean floor from an earthquake, volcanic eruption or landslide. Unlike a surface wave, a tsunami is a whole column of water that reaches from the sea floor up to the surface. It can race across oceans for thousands of miles at speeds of up to 496 miles (800 km) per hour—as fast as a jet plane. Such a giant wave might stretch for hundreds of miles from crest to crest and yet remain unnoticed as it passes under ships. A sharp rise in the ocean floor near a coastline acts as a brake at the bottom of the wave and makes it stop and rush upwards in a towering wall of water that crashes onto land. The power of the wave batters and floods the coast, causing enormous damage and loss of life. Tsunamis occur most frequently in the Pacific area.

AS FAST AS A JET PLANE
The arcs of a tsunami, triggered by an earthquake in Alaska, spread quickly across the Pacific region. Seismic sea-wave detectors are in place throughout the Pacific area to measure the travel time of tsunamis and to warn populations in danger.

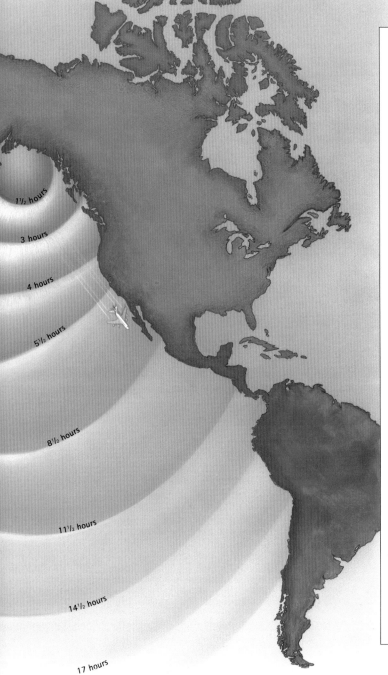

Q: How do tsunamis differ from normal ocean waves?

GIANT SEA WAVES

In 1992 a mild earthquake, barely noticed, hit San Juan del Sur in Nicaragua. Minutes later the peaceful harbor was drained dry as if someone had pulled a giant bath plug and let the water out. Amazed at the sight, people flocked to the harbor to look. As they stared, a giant tsunami rushed in and swept people and buildings far out to sea. This three-part illustration is an example of how the water is drained in a harbor, then builds up speed and height before rushing back to the shore.

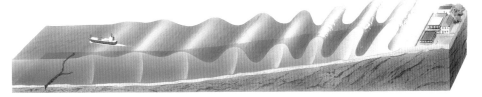

WAVE FORMATION
The speed of a tsunami depends on the depth of the ocean. The wave gets higher and higher as it moves towards the shallower water near land.

CITY IN RUINS

The earthquake struck at midday when many households were preparing their meals on hibachis—a type of open stove. As the buildings collapsed they caught fire and flames swept quickly through the city, killing thousands of people. All but one section of Tokyo was damaged by fire, and the business district was virtually destroyed.

WORLDWIDE RESPONSE
News of the devastating earthquake flashed around the world, and many countries rushed to Japan's aid with relief supplies.

• FAMOUS EARTHQUAKES •

The Great Kanto Earthquake

The Great Kanto Earthquake shook Japan on September 1, 1923. The earthquake originated beneath Sagami Bay, Yokohama, 50 miles (80 km) south of the capital, Tokyo. The power of the earthquake registered a massive 8.3 on the Richter scale, and the ground shook for nearly five minutes. A staggering 100,000 people died, and more than 300,000 buildings were destroyed. The earthquake was soon followed by a killer tsunami, which swept people and their homes far out to sea. More deaths were caused by the many fires that broke out among the paper and wood houses. These building materials had been specially chosen to make the homes safer in an earthquake, but instead they provided fuel for the raging flames. A second major tremor blasted the area 24 hours later, and minor aftershocks followed in the next few days.

Q: What are some of the after-effects of an earthquake?

Japanese Quakes

Japan is situated where the Philippine plate and the Pacific plate are subducting under the Eurasian plate. This makes Japan the site of frequent volcanic eruptions, earthquakes and tsunamis. People in Japan feel Earth tremors every few weeks. Cities such as Tokyo have disaster teams ready to jump into action. Many people have prepared emergency supply kits with food, water and medicine, and most take part in earthquake drills. In October, 1994, a major earthquake estimated at 7.9 on the Richter scale occurred in the ocean crust off Hokkaido, the northernmost island in Japan, causing many buildings to collapse (above). Three months later an earthquake estimated at 6.9 on the Richter scale shattered the city of Kobe. More than 5,000 people died.

REDUCED TO RUBBLE
So great was the force of the earthquake that the floor of Sagami Bay split. At the seaport of Yokohama, south of Tokyo, most buildings were destroyed as well as the harbor and port facilities.

• FAMOUS EARTHQUAKES •
Mexico City

An earthquake measuring 8.1 on the Richter scale struck Mexico City on September 19, 1985. As the Cocos plate subducted under the North American plate, it fractured, or cracked, 12 miles (20 km) down in the mantle. The vibrations to the ocean floor unleashed a tsunami and produced a surge of energy 1,000 times greater than an atomic bomb. The seismic shock waves travelled 217 miles (350 km) east to Mexico City. This city is built on top of a dry, sandy lake bed of soft sediment, or subsoil, which amplified the shock waves so much that many buildings collapsed. Some of the city's largest skyscrapers remained standing, but L-shaped buildings and those with large open foyers were badly damaged. The earthquake killed more than 9,000 residents, while 30,000 were injured and 95,000 were left homeless.

SHIFTING PLATES
Mexico City lies in an area where several plates meet.

ON SHAKY GROUND
Mexico City is built on the same dry lake bed as the old Aztec capital of Tenochtitlan. The area is surrounded by volcanoes and prone to earthquakes.

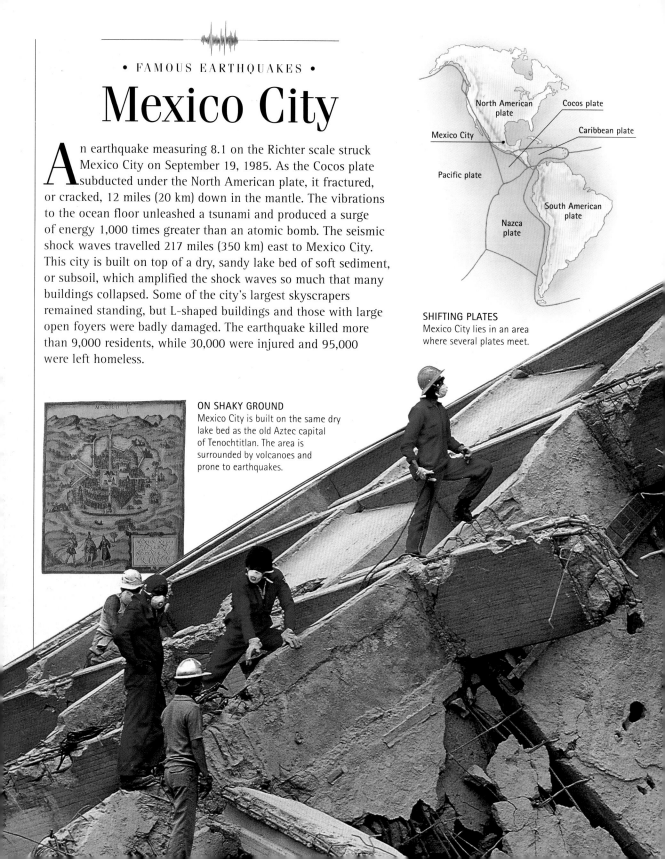

SIGNS OF LIFE
Teams of rescue workers searched carefully through the rubble of destroyed buildings looking for survivors. They listened for sounds of life in the wreckage. Rescuers worked nonstop for days after the earthquake and saved more than 4,000 lives.

A Clash of Four Plates

The Nazca, Cocos, South American and Caribbean plates meet and interact along the eastern section of the Pacific Ring of Fire. Plate movement triggers a large number of earthquakes and volcanoes in some areas of North America, Central America and South America. Pictured here is Arenal volcano in Costa Rica.

Strange but True

A four-day-old baby boy survived for nine days buried under the rubble of a hospital. Here, the miracle baby is lifted to safety by rescue workers.

Q: Why did so many buildings collapse in the Mexico City earthquake?

• FAMOUS EARTHQUAKES •

Californian Quakes

Many of the world's earthquakes occur along an edge where two lithospheric plates meet. The state of California straddles two such plates. Most of the state sits on the North American plate, which is moving very slowly. The Pacific plate, with the rest of the state, is grinding past the North American plate more quickly, and moving northwest. The grating movement of the two plates has made a weblike series of faults and cracks in the crust where earthquakes can occur. The San Andreas Fault is the most famous fault in California and slashes through the state for 682 miles (1,100 km). From time to time, the rock breaks and moves along a section of the fault, and this can trigger an earthquake. Scientists record more than 20,000 Earth tremors in California every year, although most are slight and detected only by sensitive instruments.

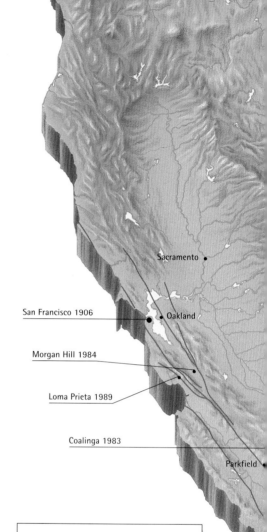

San Francisco 1906
Oakland
Morgan Hill 1984
Loma Prieta 1989
Coalinga 1983
Parkfield
Sacramento

STRANDED
Many commuters were trapped when sections of the California freeway system collapsed during the Loma Prieta earthquake in 1989.

DID YOU KNOW?
Los Angeles sits on the Pacific plate. San Francisco sits on the North American plate. Perhaps in a million years or so they will meet.

THE STREETS OF SAN FRANCISCO
A major earthquake could strike California at any time. Movement along the San Andreas fault system could cause massive destruction in cities such as San Francisco and Los Angeles. Oil refineries, chemical and atomic plants, office towers, schools, hospitals, freeways, sports arenas, amusement parks and residential areas would all be affected.

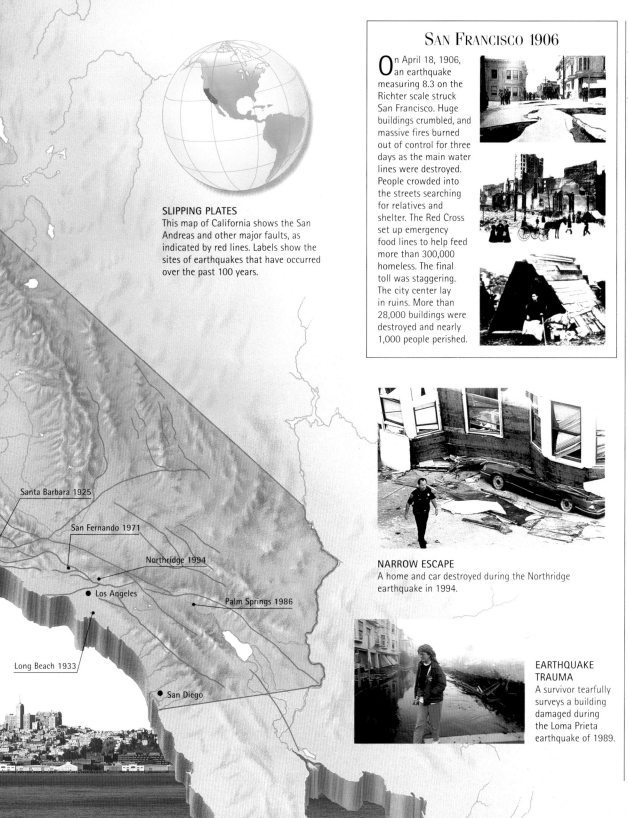

Q: Why do earthquakes occur in California?

SAN FRANCISCO 1906

On April 18, 1906, an earthquake measuring 8.3 on the Richter scale struck San Francisco. Huge buildings crumbled, and massive fires burned out of control for three days as the main water lines were destroyed. People crowded into the streets searching for relatives and shelter. The Red Cross set up emergency food lines to help feed more than 300,000 homeless. The final toll was staggering. The city center lay in ruins. More than 28,000 buildings were destroyed and nearly 1,000 people perished.

SLIPPING PLATES
This map of California shows the San Andreas and other major faults, as indicated by red lines. Labels show the sites of earthquakes that have occurred over the past 100 years.

NARROW ESCAPE
A home and car destroyed during the Northridge earthquake in 1994.

EARTHQUAKE TRAUMA
A survivor tearfully surveys a building damaged during the Loma Prieta earthquake of 1989.

Weather

- When are raindrops colored red?
- What are the strongest winds on Earth?
- Why would a tropical beach feel hotter than the Sahara Desert?

• OUR WEATHER •

What is Weather?

The weather affects all things on Earth. It helps to shape our landscapes and provide our food supplies. The weather influences the way we live, where we live, what we wear, the running of transportation systems and even how we feel. Extreme weather can bring storms that destroy homes, or droughts that can ruin crops. But what exactly is weather? It is the conditions that exist in the air around us at any one time: the temperature and pressure of the atmosphere, the amount of moisture it holds, and the presence or absence of wind and clouds. The weather is very hard to predict, and it can be incredibly diverse locally. One side of a mountain, for example, may be buffeted by high winds, while the other side may have no wind at all.

BURNING UP
Intense summer heat may help to ignite trees and cause dramatic fires. If these fires are fanned by high winds, they can move quickly through brushland.

BLOWN AWAY
Some of the strongest winds are associated with hurricanes. These huge tropical storms bring heavy rain and high winds.

SNOW FALLS
In cold weather, heavy falls of snow can make it difficult to move around.

FOOD PRODUCTION

Rain is very important to the production of food. When there is plenty of rain, crops grow and give good yields, and animals have plenty of water. But if it does not rain, even for just a few weeks, the effects can be disastrous. The soil can become parched and crops may wither and die. If this happens, grazing animals do not have enough food to eat and water supplies start to stagnate.

Q: How does the weather affect our food supplies?

UNDER THE SUN
Vast areas become parched and dry when it does not rain for months on end.

RAINBOW
A mixture of sunlight and rain often creates this colorful sight in the sky.

FLOOD
Storm clouds bring heavy rain. Rivers can burst their banks and flood low-lying land, washing away buildings and crops.

Discover more in Global Warming

• OUR WEATHER •

The Weather Engine

The sun fuels the world's weather. The surface of the Earth is warmed by sunlight. The tropics are heated most intensely, while the two poles receive the least heat. Only half the energy coming from the sun to the Earth is absorbed by the Earth's surface. The other half is reflected back into space or absorbed into the atmosphere. Different surfaces reflect varying amounts of heat. Bright white snow can reflect 90 percent of the sun's energy, so very little heat remains. The dark green tropical rainforests, however, absorb a large amount of energy. Temperatures on land change more than those in the oceans. These differences generate pressure patterns that cause winds to blow. They also set in motion the vast circulation of the atmosphere, which produces the world's weather and climate.

HEAT FROM THE SUN
Because the Earth is a sphere, air is warmed more at the equator than at the poles. Rays of sunlight contain certain amounts of energy. Depending on the season, rays will fall on small circular areas near the equator. At the poles, however, the rays are spread over a wider area because they hit the Earth at an angle. If the heat at the equator was not distributed by wind and water, the equator would get hotter and hotter.

DID YOU KNOW?
The amount of solar energy that reaches the Earth's atmosphere every 24 hours is similar to the amount of energy that would be released by 200 million electric power stations during the same period of time.

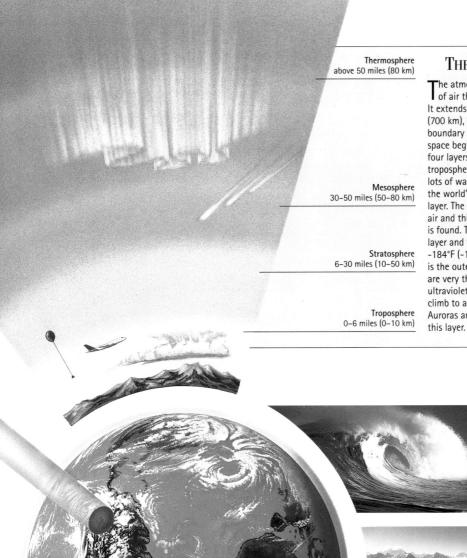

Q: How does the sun fuel the weather engine?

Thermosphere
above 50 miles (80 km)

Mesosphere
30–50 miles (50–80 km)

Stratosphere
6–30 miles (10–50 km)

Troposphere
0–6 miles (0–10 km)

THE ATMOSPHERE

The atmosphere is an envelope of air that surrounds the Earth. It extends to a height of 434 miles (700 km), but there is no clear boundary as to where it ends and space begins. The atmosphere has four layers. The lowest layer is the troposphere where the air contains lots of water vapor and dust. Most of the world's weather occurs in this layer. The stratosphere has dry, warm air and this is where the ozone layer is found. The mesosphere is a colder layer and temperatures can fall to -184°F (-120°C). The thermosphere is the outer layer and the gases here are very thin. As the gases absorb ultraviolet light, the temperatures climb to as high as 3632°F (2000°C). Auroras and meteors are seen in this layer.

MOVING WATER
The world's oceans greatly affect weather and climate. Water evaporates, which causes clouds and rain, and currents move heat from the equator to the poles.

UPS AND DOWNS
The shape of the Earth influences the weather. Mountains can deflect the wind and rain. The Himalayan mountain system has a major effect on the summer monsoon in Asia.

KEEPING CONTROL
The polar ice caps act as thermostats for the world's weather and climate. The ice and snow reflect much of the sun's energy, and any change in the area of the ice caps can affect global temperatures.

Discover more in Winds and Currents

• THE DAILY WEATHER •

Temperature and Humidity

If you were standing in the Sahara Desert, the air would feel hot and dry. If you were lying on a tropical beach, it would feel hotter, even though the temperature may be the same in both places. The reason for this difference is humidity—the amount of moisture or water vapor in the air. Humans can only tolerate a certain range of temperature and humidity. Sweating helps to keep the body cool. But if the air is very humid, water does not evaporate so easily and sweat remains on the skin. This can be uncomfortable and makes you feel hotter. Humidity is measured with an instrument called a hygrometer. A simple hygrometer uses two thermometers: one has a bulb that is surrounded by a wet cloth, while the other is dry. If the air is very dry, the "wet bulb" is cooled rapidly by evaporation. But if the air is very humid, little evaporation occurs and the reading of the two thermometers is almost the same.

GROWING CONDITIONS
A desert and a rainforest often have similar air temperatures, but lush vegetation can only grow in the rainforest because the air is very moist, or humid. The heat of the desert, however, is very dry. This lack of moisture means that few plants can survive.

KEEPING COOL
In hot weather, or when we exercise, our body temperature may rise above 98.6°F (37°C). Special glands on the surface of the skin release sweat. The evaporation of this watery fluid cools our bodies.

MEASURING THE TEMPERATURE

In 1714, the German scientist Gabriel Fahrenheit invented the temperature scale. The zero point was based on the lowest point to which the mercury fell during the winter in Germany. The freezing point of water was 32°F, while the boiling point was 212°F. In 1742, the Swedish astronomer Anders Celsius proposed an alternative scale. He suggested making the freezing point of water 0°C and the boiling point of water 100°C. This scale was very useful for scientific work, and it is used more widely than Fahrenheit's scale.

Anders Celsius

Hot and dry
This hygrometer shows a humidity reading of 20 percent.

A FROZEN DESERT
Antarctica is the coldest place on Earth. Its freezing air tends to be extremely dry because very cold air holds only a small amount of water vapor.

Fronts
When masses of air of different temperatures meet, the warm air is forced to rise. Clouds appear as the warm air rises and this is called a warm front.

CLOUD FORMATION
Clouds can form in many ways but they all involve warm moist air coming into contact with cooler air.

Current collision
When currents of air collide, they force each other upward.

• THE DAILY WEATHER •

What are Clouds?

Clouds are masses of water droplets and ice crystals that float in the sky. They are formed by a process that begins when warm moist air rises. As the warm air cools, it is unable to hold water vapor. Some of the water vapor condenses around dust particles and forms minute water droplets. These tiny drops make up clouds. The sky can be covered with a blanket of cloud that is formed when a mass of warm air rises above cooler air and causes the water vapor to condense. Clouds also form when warm air is forced to rise over mountains, or when warm air blows over a colder surface such as cool water. On hot days, storm clouds appear when warm moist air rises and then cools rapidly. Clouds appear to be white because the water droplets reflect light. As a cloud becomes thicker and heavier with droplets, it darkens because light cannot pass through it.

AN AERIAL VIEW
Seen from above, the tops of clouds often look like a white blanket or a field of snow.

Water droplets
Millions of microscopic droplets of water are needed to make one drop of rain.

THE WATER CYCLE

Water covers more than 70 percent of the Earth's surface. Warmth from the sun causes some of the water to evaporate from the surface of oceans, lakes and rivers, as well as from plants. This water vapor rises and cools, and then condenses back into water to form clouds. The water droplets fall as rain or snow, which runs into rivers and lakes and can sometimes soak into underground layers of rock. Eventually, the water returns to the oceans to complete the cycle.

Convection
On a sunny day, some patches of ground warm up more quickly than others. Bubbles of warm air form over these spots and rise into the sky. As the bubbles rise, they expand and cool. The water condenses and forms a cumulus cloud that is shaped like a dome.

In the clouds
When warm air meets a mountain range, it is forced upward. The warm air cools and banks of clouds form around the mountain peaks.

CLOUD BANK
Banks of clouds are often seen lying over mountain ranges.

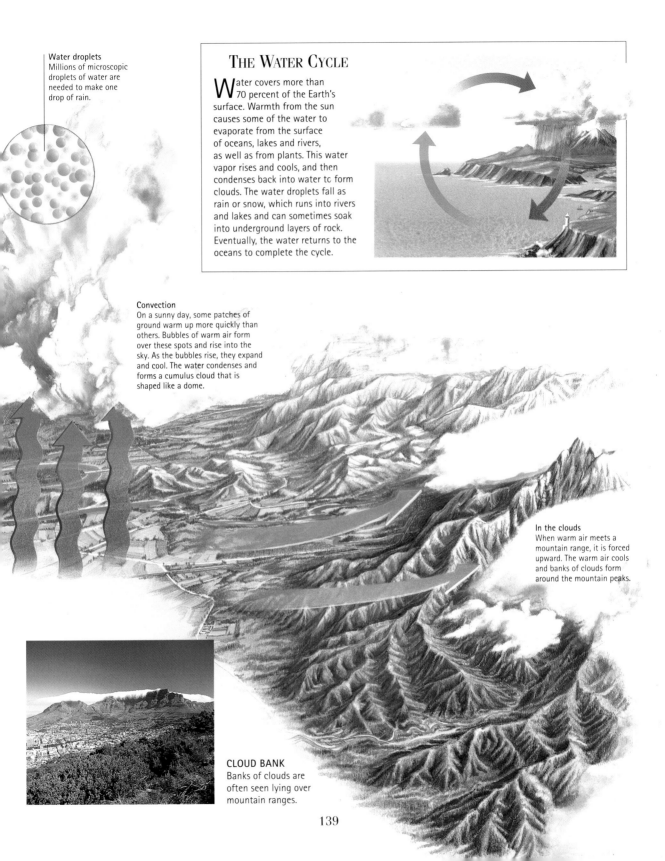

Cirrus
32,800 ft (10,000 m)
Cirrocumulus
Cirrostratus
19,700 ft (6,000 m)

HIGH IN THE SKY
The three main cloud groups are based on height. High clouds are more than 19,700 ft (6,000 m); middle clouds are 6,600–19,700 ft (2,000–6,000 m); and low clouds are under 6,600 ft (2,000 m). Cumulonimbus clouds tower higher than 32,800 ft (10,000 m).

Altocumulus

Altostratus

Cumulonimbus

6,600 ft (2,000 m)

Stratocumulus

Stratus

Cumulus

Nimbostratus

• THE DAILY WEATHER •

Types of Cloud

No two clouds are exactly the same. Although they vary in shape and size, they can be divided broadly into two similar types: heaped, fluffy clouds; and layered clouds. Heaped clouds are formed when pockets of warm air drift upward, while layered clouds are created by moist air moving horizontally between cooler layers. Clouds are usually grouped according to how high they are above the ground. It is important to identify different types of cloud because they give us information about the weather. White, puffy cumulus clouds, for example, are associated with warm sunny days. High cirrus clouds mark the approach of a weather front (an advancing mass of warm or cold air). Cirrus clouds may be followed by lower altostratus clouds and low stratus rain clouds, which cover the entire sky in a solid gray sheet.

BUBBLING SKIES
These pendulous clouds, called mammatus clouds, form below the anvil of a thundercloud. They are frequently seen with storms that produce tornadoes.

FLYING SAUCERS
These lenticular, or lens-shaped, clouds have been mistaken for flying saucers. They usually form in bands on the sheltered side of mountain ranges.

MIXED SKIES
At least five cloud types are visible in this busy sky. In the background, a huge pale cumulonimbus thundercloud fills the sky. Along the lower edge, dark rain-bearing stratus clouds underlie paler, fluffier cumulus clouds. The strong, dark streaks at the middle level are altostratus, with altocumulus above and below.

Did You Know?
Cumulonimbus are the largest clouds and can tower 11 miles (18 km) into the sky. These clouds may contain up to 110,000 tons (100,000 tonnes) of water.

Vapor Trails
Some aircraft leave white vapor trails as they fly across a clear blue sky. These "artificial clouds" are caused when the hot exhaust gases from the jet engines mix with the surrounding cold air and cool rapidly. Water vapor within the exhaust freezes and forms a trail of ice crystals.

• THE DAILY WEATHER •

Thunder and Lightning

LIGHTNING STRIKE
Lightning usually strikes the highest point, such as a tall building or an isolated tree, so it is dangerous to shelter beneath a tree during a thunderstorm. You would be much safer in a car.

On a hot, humid summer day, strong rising convection currents of warm air form cumulus clouds that soon grow into a towering cumulonimbus cloud, or thundercloud. These black clouds are accompanied by strong winds, heavy rain, lightning and thunder, and often produce spectacular summer storms. Most lightning occurs in cumulonimbus clouds because they contain violent currents of air and a plentiful supply of super-cooled droplets of water. The intense heating of the air by lightning causes the air to expand at supersonic speed and produces a clap of thunder. Lightning and thunder occur at the same time, but as light travels faster than sound, we see the flash before we hear the thunder. We can tell how far away a storm is by timing the interval between the flash and the thunder. A three-second interval represents a distance of $^6/_{10}$ mile (1 km).

RIBBON LIGHTNING
Strong winds may cause lightning to move and give a ribbonlike effect.

CLOUD TO GROUND
Lightning can form when there is a build up of negative charges at the bottom of a cloud and the ground below is positively charged.

CLOUD TO CLOUD
Lightning may occur between a negatively charged cloud and a nearby cloud that is positively charged.

INSIDE A CLOUD
Most lightning forms within a cloud, when there is a discharge between a positive and negative charge.

DID YOU KNOW?
Lightning strikes the Earth as frequently as 100 times every second. These strikes are produced by the 2,000 thunderstorms that rage around the world at any one moment in time.

BALLS OF FIRE
Sometimes, lightning appears as a fiery ball. Some balls disappear quietly, while others explode. Some have even appeared to chase people! Fortunately, ball lightning is very unusual and seldom causes harm.

Life of a Thundercloud

Strong currents build up within a developing thundercloud. These turbulent currents cause ice crystals within the cloud to continually rise and fall. The ice crystals become heavier and heavier as fresh layers of ice are added to the crystals. Toward the end of the thundercloud's life, the ice crystals are so heavy that the air currents cannot keep them airborne. Instead, they fall to the ground as ice and rain. This marks the end of the thunderstorm and the cloud begins to break up.

A dying thunderstorm

• THE DAILY WEATHER •
Rain, Hail and Snow

The water or ice that falls from a cloud is called precipitation. This may be in the form of rain, drizzle, sleet, snow or hail. The conditions within a cloud, and the temperature outside it, determine the type of precipitation that falls. One important factor is whether the cloud is high enough above the ground for the water droplets in it to turn into ice. The height at which this occurs is called the freezing level. It can be as little as 1,000 ft (300 m) or as high as 16,000 ft (5,000 m) above the ground. Snow falls from low and very cold clouds when the air temperature is around freezing, so the ice crystals can reach the ground without melting. If snow falls into air that is just above freezing, some of the crystals melt and produce a mixture of rain and snow called sleet. Dark cumulonimbus clouds bring thunderstorms, which may be accompanied by hail. A blanket of thin nimbostratus clouds produces a steady stream of rain, while low stratus clouds bring drizzle.

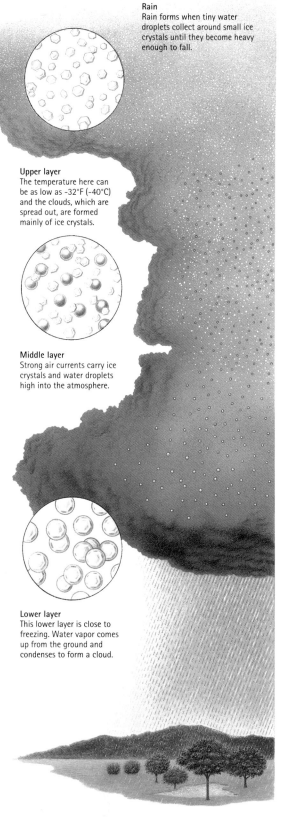

Rain
Rain forms when tiny water droplets collect around small ice crystals until they become heavy enough to fall.

Upper layer
The temperature here can be as low as -32°F (-40°C) and the clouds, which are spread out, are formed mainly of ice crystals.

Middle layer
Strong air currents carry ice crystals and water droplets high into the atmosphere.

Lower layer
This lower layer is close to freezing. Water vapor comes up from the ground and condenses to form a cloud.

RAINY DAYS
A low blanket of gray clouds often brings a steady downpour that can last for an hour or more. However, drizzle and light rain can last for much longer.

Hail
Hail forms around small ice crystals. As strong air currents circulate repeatedly and cause layers of ice to build up around the crystals, hailstones become larger.

Snow
Snow forms if the freezing level is below a height of 1,000 ft (300 m) above the ground, and the ice crystals do not have time to melt before they reach the ground.

STRANGE BUT TRUE
In 1953, hailstones as large as golf balls fell in Alberta, Canada. They covered an area 140 miles (225 km) by 5 miles (8 km) and killed thousands of birds.

HAIL STORM
Hail can cause great damage everywhere. This crop was ruined completely by these large hailstones.

THE SHAPE OF A SNOWFLAKE
Snowflakes are loose clusters of ice crystals usually with a flat, six-sided (hexagonal) shape. The exact shape of a snowflake depends on the temperature of the air, and no two snowflakes are the same. At very low temperatures, small needlelike crystals form, while at temperatures nearer to freezing, larger branching shapes are more common. The amount of water vapor in the air is important: at lower temperatures there is less water vapor and the crystals that form tend to be smaller.

Discover more in What are Clouds?

Q: Why are snowflakes different shapes?

• THE DAILY WEATHER •

Fog, Frost and Ice

COLD AND FROSTY MORNINGS
On clear nights, heat from the ground escapes back into space. Temperatures near the ground may be low enough for ice to form. Hoar frost is a coating of tiny ice crystals. If there is an icy wind, the temperature drops further and a thick coating of ice appears on exposed surfaces. This is called rime frost.

Clear nights with low ground temperatures often bring fog and frost, especially around dawn. The clear skies allow heat to radiate into space, and the temperature drops toward freezing. Moisture in the air condenses near the ground to form low-lying fog or mist. In the morning, the sun heats the air and the fog disperses. If the temperature is low enough, moisture in the air freezes and coats the ground, plants and other surfaces with a thin layer of frozen dew or frost. In long periods of cold, the surfaces of ponds and lakes freeze over, while dripping water freezes into icicles. Smog is a visible form of air pollution that often occurs in cities. It can be caused by the smoke from cars and huge industrial chimneys. Smog can affect everyone, but especially the health of the young, the old and those with lung problems.

ICY FEATHERS
On winter mornings, windows are sometimes covered with beautiful patterns of ice crystals. This happens when moisture comes into contact with the cold window and is cooled to the freezing point.

Types of Fog

Different conditions form different types of fog. An advection fog is produced when warm, moist air blows over a cooler surface. Radiation, or ground, fog occurs on clear nights, especially in river valleys where the ground cools quickly and moisture in the air condenses. Upslope fog is created when air is forced to rise up the side of a hill and cools, which causes moisture in the air to condense. Frontal fogs form when weather fronts, especially warm fronts, pass through a cooler area.

Q: How is smog formed?

A WALL OF ICE
Sometimes it is so cold that ocean spray freezes, and a wall of ice forms along the shoreline.

Did You Know?

Before the invention of refrigerators, ice was cut from ponds in winter and stored in an ice house. The ice house was a pit, packed with layers of ice and straw and covered with an insulated roof. Cold air sinks, so the pit remained very cold and the ice lasted right through the summer.

GHOSTLY SHADOWS
Sometimes, the sun projects enlarged shadows of mountain climbers onto low-lying clouds. This creates a ghostly effect called a Brocken Specter.

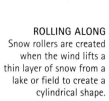

ROLLING ALONG
Snow rollers are created when the wind lifts a thin layer of snow from a lake or field to create a cylindrical shape.

• THE DAILY WEATHER •

Weather Wonders

COLOR OF LIGHT
Sunlight is called white light, but it is really a mixture of different colors. Rays of sunlight are bent as they pass into the raindrop. They reflect off the back of the drop and bend again as they leave.

You can often see strange effects in the sky. When sunlight hits ice crystals or water droplets, some of the colors of light are reflected. This creates phenomena such as rainbows, sundogs and glories. The multicolored arc of a rainbow stretching across the sky directly opposite the sun is one of the most colorful sights in the sky. Rainbows occur during isolated showers or thunderstorms, when falling rain is illuminated by sunlight. Sundogs, or parhelia, are small suns that appear on either side of the sun. They are created when ice crystals bend light in high clouds. When water droplets in the clouds reflect sunlight back toward the sun, alternating red and blue rings—called glories—are visible beneath the sun. Colored rain is an unusual weather wonder. Red sand carried from desert areas by the wind can cause red rain, while soot in the air can result in black rain. Yellow rain, caused by pollen, is very rare.

SUNDOGS
These are images that appear to the left and right of the sun at the same height above the horizon. They are called sundogs because they often have long, white tails that point away from the sun.

Q: What is a rainbow?

SEEING DOUBLE

Rainbows are caused by the reflection of sunlight in millions of raindrops. The sun must be behind you and fairly low for a rainbow to be visible in the sky. This is why rainbows are never seen in the middle of the day. Sometimes, if the light has been reflected twice inside each raindrop, a second fainter rainbow can be seen about 9 degrees outside the first. The colors are reversed in order, with red on the inside.

FALLING FROM THE SKY

An old saying, "it's raining cats and dogs," may not be as strange as it sounds. Strong storm winds can suck up quite large objects. In October 1968, during a heavy storm in Acapulco, Mexico, maggots reportedly rained down on boats in the harbor. Stories of fish and frogs falling from the sky have been told around the world for centuries. The record books also tell of worms, snails and even snakes raining on surprised people.

• WEATHER FORECASTING •

Weather Watch

The Earth's atmosphere is a massive and constantly moving weather machine. Weather forecasters need to gather information about the atmosphere from all around the world, both at the Earth's surface and at heights of up to 2,480 miles (4,000 km). There are thousands of weather stations on land and at sea recording these changes in the atmosphere—the changes we call weather. These observations are backed up by balloons and aircraft that take atmospheric readings. In some places throughout the world, automatic and manual weather stations are found in remote places. All weather stations are required to take readings in the same way and their reports are gathered together for analysis by the World Meteorological Organization. Separate national organizations obtain the information necessary for preparing the forecasts we read or hear.

DID YOU KNOW?
Rain gauges were used in India as long ago as 400 BC. Farmers would place a number of small bowls in different places to catch rain. This helped them to learn about patterns of rainfall.

REMOTE READINGS
Meteorologists need information from all around the world, so there are weather stations in some extremely remote areas. Some stations are run by trained observers, although automatic weather stations are now becoming more common.

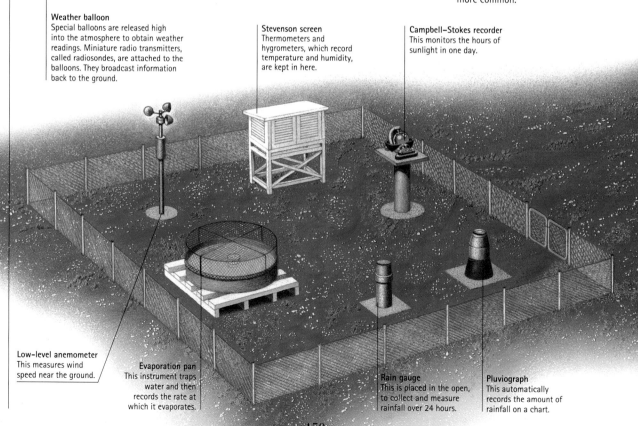

Weather balloon
Special balloons are released high into the atmosphere to obtain weather readings. Miniature radio transmitters, called radiosondes, are attached to the balloons. They broadcast information back to the ground.

Stevenson screen
Thermometers and hygrometers, which record temperature and humidity, are kept in here.

Campbell–Stokes recorder
This monitors the hours of sunlight in one day.

Low-level anemometer
This measures wind speed near the ground.

Evaporation pan
This instrument traps water and then records the rate at which it evaporates.

Rain gauge
This is placed in the open, to collect and measure rainfall over 24 hours.

Pluviograph
This automatically records the amount of rainfall on a chart.

SEA SPY
Conditions at sea are monitored by specially equipped ships and remote weather buoys. These buoys are towed to positions away from shipping lanes and anchored to the sea bed. It is important to watch the weather far out at sea, because severe storms and hurricanes form there.

EYE IN THE SKY

Meteorology, or the study of weather, changed dramatically in 1960, when the first weather satellites were launched into space. These satellites scan huge areas of the Earth and send back a range of measurements, as well as images of cloud cover and other weather conditions. This information enables meteorologists to plot the development and course of major events such as hurricanes, and predict the weather more accurately. By using satellite sensors that are sensitive to heat and light, meteorologists can also obtain information about the temperature of different types of clouds and the surface of the land and sea.

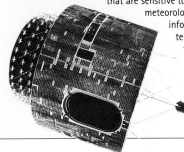

Meteorological satellite

Q: Where can weather observation stations be found?

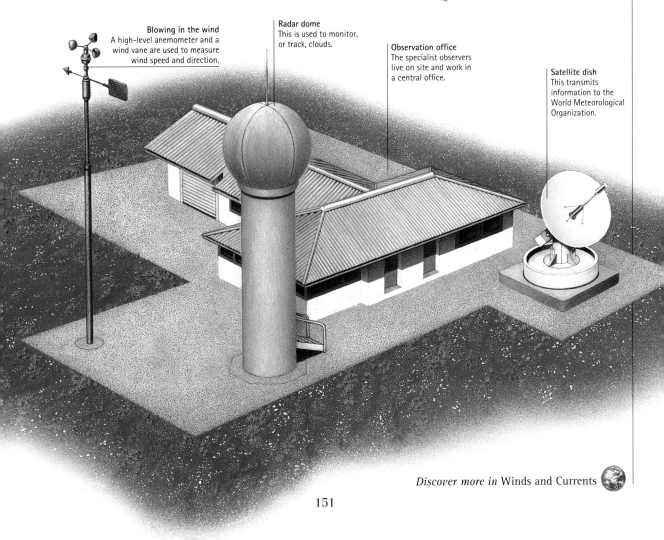

Blowing in the wind
A high-level anemometer and a wind vane are used to measure wind speed and direction.

Radar dome
This is used to monitor, or track, clouds.

Observation office
The specialist observers live on site and work in a central office.

Satellite dish
This transmits information to the World Meteorological Organization.

Discover more in Winds and Currents

• WEATHER FORECASTING •

Forecasting

Every minute of the day and night, weather recordings from observation stations, ships, planes and satellites are received by meteorological offices all around the world. These recordings form a vast databank, from which meteorologists gather information. The system that enables this huge exchange of information is the Global Telecommunications System (GTS). The data is fed into powerful computers that enable meteorologists to prepare weather maps known as synoptic charts. Meteorologists study these charts very carefully and compare them to previous charts before they produce a weather forecast. The weather presenters we see on television use synoptic charts to prepare weather maps with simple symbols such as rain clouds and yellow suns. Weather forecasts are 85 percent accurate for the next few days. However, it is far more difficult to predict the weather for more than a week ahead.

ON THE MAP
Synoptic charts contain a wealth of information including air pressure, wind speed and direction, cloud cover, temperature and humidity. The most noticeable features on the charts are isobars. These are lines that join places of equal air pressure and are measured in hectapascals. Isobars that are close together, as they are on this map, show an area of low pressure. This usually brings wind and rain.

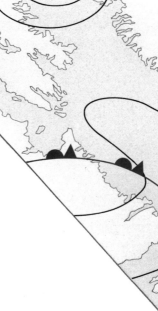

ON THE DRAWING BOARD
Meteorologists spend many hours preparing synoptic charts. All the different information has to be carefully plotted on the chart.

KEY TO SYMBOLS

wind symbols
- light
- high
- gale force

cloud symbols
- clear sky
- partly cloudy
- cloudy

weather symbols
- cold front
- warm front
- occluded front
- snow
- fog
- rain
- sleet

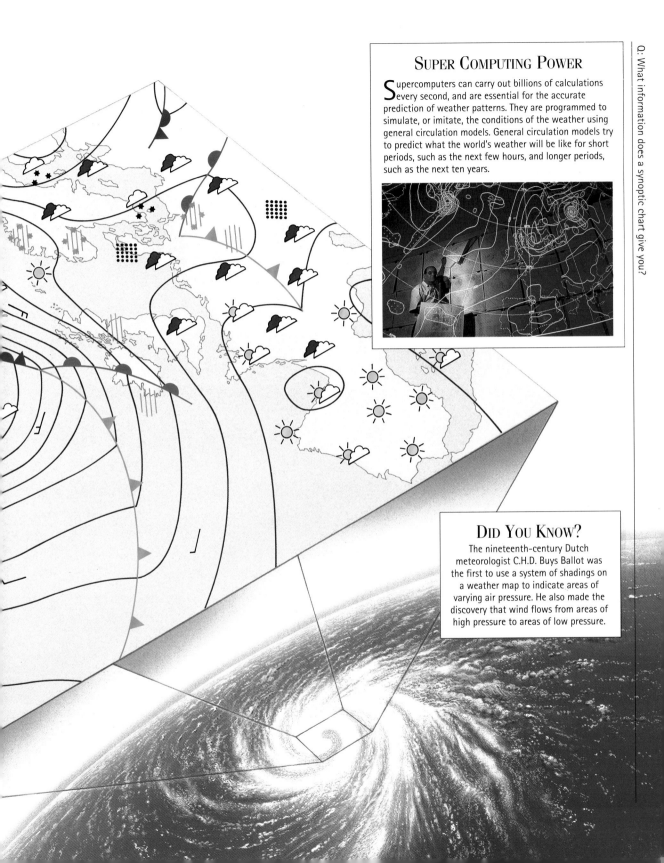

Super Computing Power

Supercomputers can carry out billions of calculations every second, and are essential for the accurate prediction of weather patterns. They are programmed to simulate, or imitate, the conditions of the weather using general circulation models. General circulation models try to predict what the world's weather will be like for short periods, such as the next few hours, and longer periods, such as the next ten years.

Did You Know?

The nineteenth-century Dutch meteorologist C.H.D. Buys Ballot was the first to use a system of shadings on a weather map to indicate areas of varying air pressure. He also made the discovery that wind flows from areas of high pressure to areas of low pressure.

Q: What information does a synoptic chart give you?

• CLIMATE •

Winds and Currents

Winds constantly circle the Earth. They bring rain and influence temperatures. The polar easterlies, prevailing westerlies and trade winds are called prevailing winds because they cover large sections of the Earth. Small, circular wind flows are called cells. Jet streams move air between these cells high in the atmosphere and at very high speeds. Sailors have known about the patterns of wind for centuries, and many of the winds, and areas near them, were named by early sailors. The horse latitudes, for example, occur in the Atlantic Ocean. When sailing ships with cargoes of horses for the New World encountered calm, hot weather in this area, many of the horses died. The Atlantic trade winds blew trading ships between Europe and the New World, while the narrow, windless area around the equator, called the doldrums, has frustrated sailors through time. Ocean currents follow the direction of the prevailing winds and affect both the climate of the world and our daily weather.

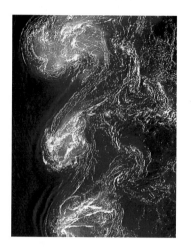

COMPUTER CURRENTS
Information about the oceans, including wind and temperature, are fed into computers that produce maps of ocean currents. This map shows an Antarctic current running across the bottom. The red areas indicate fast-flowing water while the blue areas are slow currents.

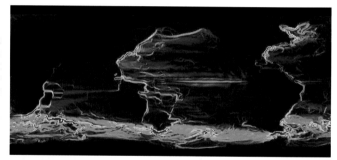

OCEAN CURRENTS

Ocean currents follow the direction of the prevailing winds. In each ocean there is a roughly circular movement of water called a gyre. Near the equator, the currents are blown toward the west, but at the poles the currents flow eastward. In this diagram it is possible to see the warm Gulf Stream. It runs up the coast of the eastern United States and then turns eastward across the Atlantic Ocean to Northern Europe. The Gulf Stream brings mild weather to parts of Northern Europe that would otherwise be much cooler.

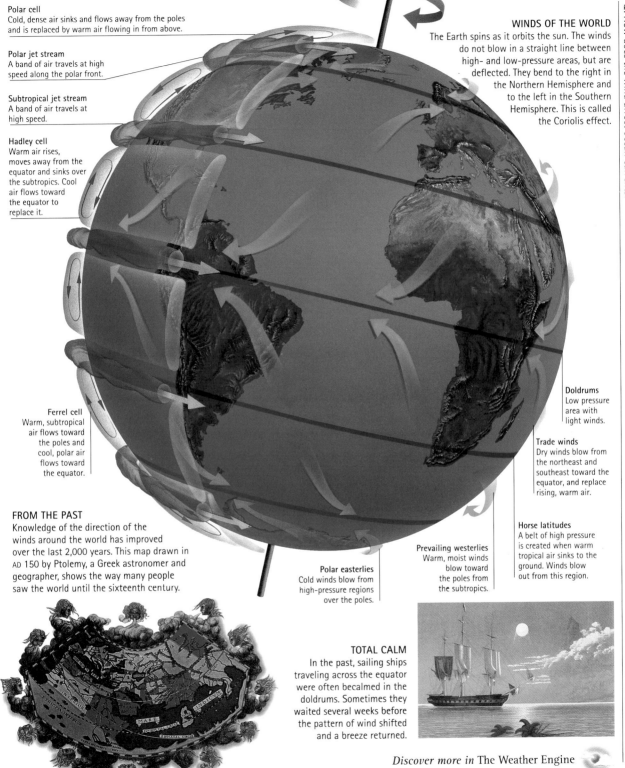

Polar cell
Cold, dense air sinks and flows away from the poles and is replaced by warm air flowing in from above.

Polar jet stream
A band of air travels at high speed along the polar front.

Subtropical jet stream
A band of air travels at high speed.

Hadley cell
Warm air rises, moves away from the equator and sinks over the subtropics. Cool air flows toward the equator to replace it.

Ferrel cell
Warm, subtropical air flows toward the poles and cool, polar air flows toward the equator.

WINDS OF THE WORLD
The Earth spins as it orbits the sun. The winds do not blow in a straight line between high- and low-pressure areas, but are deflected. They bend to the right in the Northern Hemisphere and to the left in the Southern Hemisphere. This is called the Coriolis effect.

Doldrums
Low pressure area with light winds.

Trade winds
Dry winds blow from the northeast and southeast toward the equator, and replace rising, warm air.

Horse latitudes
A belt of high pressure is created when warm tropical air sinks to the ground. Winds blow out from this region.

Prevailing westerlies
Warm, moist winds blow toward the poles from the subtropics.

Polar easterlies
Cold winds blow from high-pressure regions over the poles.

FROM THE PAST
Knowledge of the direction of the winds around the world has improved over the last 2,000 years. This map drawn in AD 150 by Ptolemy, a Greek astronomer and geographer, shows the way many people saw the world until the sixteenth century.

TOTAL CALM
In the past, sailing ships traveling across the equator were often becalmed in the doldrums. Sometimes they waited several weeks before the pattern of wind shifted and a breeze returned.

Discover more in The Weather Engine

Q: How does the wind affect ocean currents?

• CLIMATE •

World Climate

As the Earth is a sphere, the equator receives more heat from the sun's rays than the poles. Farther from the equator, the sun is weaker and less heat reaches the Earth's surface. The surface of the Earth is not heated equally, and this results in a pattern of winds moving the air around constantly. The intense heat of the equatorial sun causes warm, moisture-laden air to rise into the atmosphere. As this air cools, the moisture condenses and falls as rain. Warm air moves away from the equator and eventually sinks to the ground, which helps to form deserts such as the Sahara. Cool air is drawn back toward the equator to replace the rising, warm air. This sets up the circulations of air that produce the world's climates. The distribution of land and sea on the Earth, and the presence of mountain ranges, also affect climate. Coastal regions have milder climates than areas in the middle of a continent. Ocean currents influence climate as well. Northwest Europe has a mild climate because the warm waters of the Gulf Stream pass nearby.

WORLD CLIMATE REGIONS
This map shows the world's major climate zones. Climate is the typical weather of a region, based on average weather conditions over a period of at least 30 years.

THE SEASONS

Regular changes in weather patterns during the year are called seasons. In many parts of the world there are four seasons—spring, summer, fall and winter—while in other areas there are only two—a wet and a dry season. The Earth is tilted at an angle as it circles the sun. For six months of the year, the Northern Hemisphere is tilted toward the sun. It has long, warm, summer days while the Southern Hemisphere has short, cool winter days. For the next six months, the Northern Hemisphere is tilted away from the sun. This part of the world has winter while the Southern Hemisphere enjoys summer.

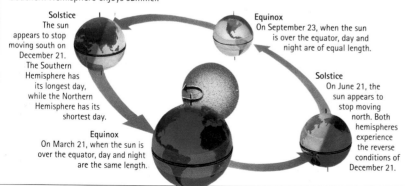

Solstice
The sun appears to stop moving south on December 21. The Southern Hemisphere has its longest day, while the Northern Hemisphere has its shortest day.

Equinox
On March 21, when the sun is over the equator, day and night are the same length.

Equinox
On September 23, when the sun is over the equator, day and night are of equal length.

Solstice
On June 21, the sun appears to stop moving north. Both hemispheres experience the reverse conditions of December 21.

POLAR ZONES
These are the coldest parts of the world. Winter temperatures fall below -58°F (-50°C).

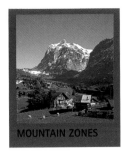
MOUNTAIN ZONES
These high altitudes have cold climates.

TEMPERATE ZONES
These have moderate temperature ranges.

TROPICAL ZONES
These have average monthly temperatures of 80°F (27°C), and high rainfall.

DESERT ZONES
Temperatures here may range from more than 104°F (40°C) in the day to freezing at night.

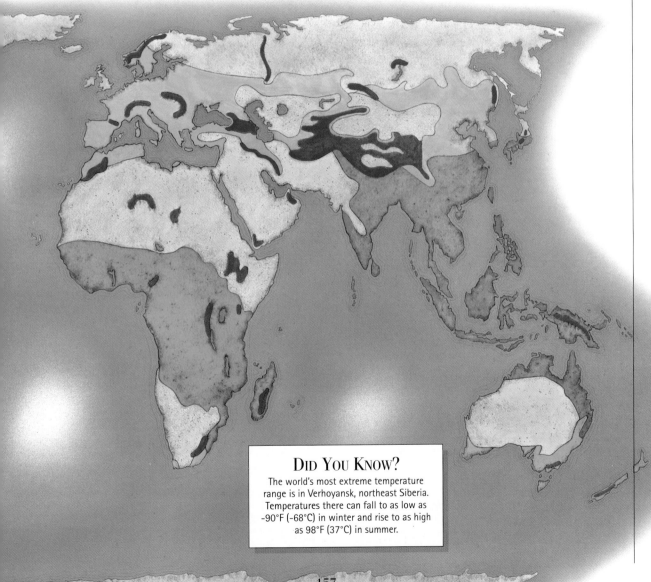

DID YOU KNOW?
The world's most extreme temperature range is in Verhoyansk, northeast Siberia. Temperatures there can fall to as low as -90°F (-68°C) in winter and rise to as high as 98°F (37°C) in summer.

• CLIMATE •

Polar Zones

Climates near the North and South poles are characterized by freezing temperatures and permanent snow and ice. Polar summers are short and cold. The extreme climate is caused by lack of heat because the sun is weaker and the ice reflects much of the heat from the sun back into the atmosphere. For six months of the year, the Arctic experiences winter as the North Pole is tilted away from the sun. At the same time, Antarctica, the continent around the South Pole, enjoys a brief summer. Temperatures rise to freezing, or just above, near the coast. The pack ice drifts northward and melts in the warmer waters. Winter in the Antarctic, however, is severe. Antarctica doubles in size as the sea freezes over, and pack ice extends for hundreds of miles around the continent. Frequent blizzards and fierce winds rage across the icy surface.

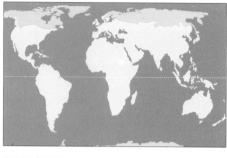

POLES APART
The Arctic is a frozen sea surrounding the North Pole, while the Antarctic is a frozen continent around the South Pole.

THE DEEP FREEZE
The Antarctic is covered in ice and snow, and the climate is bleak and hostile. Even in summer, temperatures barely reach freezing point. In spite of this, many animals live in the polar regions.

COAT OF COLORS
The fur of the Arctic fox changes color during the year. In winter, it turns from smoky gray to white to camouflage the fox against the snow.

ARCTIC DWELLERS
The Inuit (Eskimos), who live in the Arctic, have adapted well to the extreme climatic conditions.

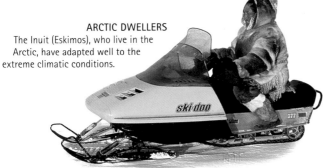

WHITE OUT

Blizzards are strong winter snowstorms. They are particularly severe in the polar zones, where they may last for weeks at a time. Snow falls on more than 150 days of the year and is swept into huge piles by the wind. Winds are equally severe and reach speeds of more than 186 miles (300 km) per hour. The average winter temperatures plummet to -76°F (-60°C). In these extreme temperatures, unprotected human skin will freeze in seconds. People need layers of warm clothing and protective shelters to survive this bitter cold.

Q: What causes the extreme polar climate?

NEPAL
Nepal lies in the Himalayas. There is very little flat land for villages, so houses are scattered. The warm, sunny, south-facing slopes are used for farming. The north-facing slopes are usually forested. Because the steep slopes are difficult to farm, they have been gradually terraced to provide many small, level fields for farming. The higher pastures extend up to the permanent snow line.

CLINGING TO THE GROUND
The stems of alpine plants hug the ground to avoid the full force of the wind. Their leaves are small and waxy to reduce water loss. Because these plants grow only on warm days, they grow very slowly.

• CLIMATE •

Mountain Zones

Each mountain has its own weather pattern. Within a mountain range there may be varying climatic conditions—the side of a mountain facing the wind may experience higher precipitation than the more sheltered side. Even the position of a rock or tree, creating a barrier to the wind or snow, can have an effect on the climatic conditions. The air temperature decreases by a few degrees for every few hundred yards rise in altitude. The air becomes thinner, the sky bluer and the sun's rays stronger. On the highest mountains, there is snow and ice all year round. Nothing can survive permanently on mountain peaks that are more than 23,000 ft (7,000 m) high, because fierce winds and low temperatures would freeze any living cells. Mountain weather is very changeable too. It can be bright and sunny, then stormy. Warm daytime temperatures may be followed by bitterly cold nights.

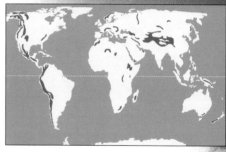

MOUNTAIN SITES
The areas highlighted above show the mountain zones around the world.

A SURE FOOTING
Mountain goats are nimble animals that can leap from rock to rock. They are found on the highest slopes, where they are safe from predators.

FLYING HIGH
Strong, gusty mountain winds make flying difficult for all but the largest birds. Eagles have broad wings that are ideal for gliding. They make good use of rising air currents to soar around the mountain peaks.

SNOW FUN
Over the last 100 years, skiing has become a popular winter sport in the mountains of Europe, North America, New Zealand, Asia and Australia. The snow-covered high mountains are used as ski runs, and ski lifts carry skiers from the valleys to the slopes. But alpine environments are slowly being damaged by all the activity in these mountain areas.

SEASONAL CHANGES
Temperate regions have distinct seasons. In spring, plants begin to produce leaves and flowers and animals start to breed. This period of growth peaks in summer. As fall approaches, deciduous trees begin to drop their leaves and many animals migrate, while others prepare to hibernate. During the winter months, snow may cover the ground for weeks at a time, and the barren landscape shows little sign of activity.

• CLIMATE •

Temperate Zones

The temperate zones of the world experience a mild, moist climate dominated by cool, moist air blowing from the poles toward the tropical zones. Large swings in temperature, the distance from the equator and the varying hours of sunlight create a changeable climate with distinct seasons. The temperate zones can be divided into three regions: warm temperate between 35 and 45 degrees latitude; cool temperate between 45 and 60 degrees latitude; and cold temperate, which is experienced by regions lying in the center of continents. The majority of people live in the temperate zones, where there is an adequate supply of water for most of the year. The world's temperate grasslands are found in these zones, where huge herds of grazing animals such as buffalo once roamed. Today, farmers keep sheep and cattle, and much of the grassland has been plowed to grow crops and grains.

AROUND THE WORLD
The temperate zones lie between 35 and 60 degrees latitude, north and south of the equator.

A CHANGE OF COLOR
Many trees have adapted to the temperate seasons. They lose their leaves during the cold months and create spectacular fall scenes such as this.

BUSY BEES
Honey bees need warmth to maintain a regular body temperature. They have adapted to the conditions in temperate zones to keep up their busy way of life.

Temperate Cities

There are differences between warm-temperate and cool-temperate climates. The warm-temperate areas receive most of their rainfall in winter and have hot, dry summers. The lack of water in summer means that the vegetation is sparse and shrubby. Cool-temperate areas have cold winters with heavy snowfall and warm, humid summers. Because rain falls all year round, vegetation is plentiful.

Winter in a warm-temperate climate

Winter in a cool-temperate climate

Q: What are the three regions within the temperate zone?

Did You Know?

The world's climate is constantly changing. Between the fifteenth and nineteenth centuries, the River Thames in London, England froze over every year. But this has not happened for the last 160 years.

Discover more in Temperature and Humidity

• CLIMATE •

Tropical Zones

Tropical zones are the warmest regions of the world. The sun is overhead for most of the year, so the climate is always hot. But there are many variations of climate within the tropical zones. Tropical wet climates are hot and humid all through the year, and have very heavy rainfall. These regions lie close to the equator and have dense tropical rainforests. Hot air, laden with moisture, rises into the atmosphere during the day. As the air cools, the water condenses to form dark clouds that bring heavy rain in the afternoons. In a tropical dry climate, a wet season is usually followed by a dry one. The wet season has heavy rain storms and hot, humid weather. The temperature can be even higher in the dry season as the days are sunny and clear. The subtropics are regions that border the tropics, and they are mostly dry.

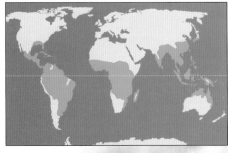

TROPICAL AND SUBTROPICAL SITES
Tropical zones lie between 30 degrees latitude north and south of the equator. Subtropical zones border these zones.

IN THE WET
When a low-pressure area develops over the land, cool, moist air from the ocean flows in. The air warms and rises as it crosses the land, and forms widespread rain clouds. The rains brought by these winds are called monsoons, and they provide 85 percent of Asia's annual rainfall.

TROPICAL DIVERSITY
In areas of South America, high temperatures and heavy rainfall support the lushest vegetation found on Earth: tropical rainforests. Many species of animal live in rainforests and feed on the fast-growing plants.

A DRY TIME
A high-pressure area over the land causes the winds to change direction. It rains out at sea and the land becomes dry.

THE SUBTROPICS

The subtropics lie to the north and south of the tropical zones. These areas do not receive as much rain as the tropics, but the temperature can be much higher. During the dry season, hot, scorching winds blow off the deserts. The ground dries and the vegetation becomes parched. As the sun moves overhead, the dry winds are replaced by hot, humid winds carrying moisture. This marks the beginning of the rainy season, which may last for several months. Zimbabwe (below) lies in a subtropical area.

• CLIMATE •

Desert Zones

Deserts cover one-seventh of the Earth's land surface. They are dry regions that on average receive less than 4 in (100 mm) of rain per year. In some deserts, rain may not fall for many years. Then, quite suddenly, a storm will break and there is a huge downpour that lasts just a few hours. The absence of moisture in the air means that clouds are rare and the skies are clear for most of the year. The land is heated by the sun and daytime temperatures soar to 104°F (40°C) and above. However, the lack of cloud cover means that much of the heat radiates back into the atmosphere at night, and the air temperature plummets to almost freezing. Although they are dry and often very windy, not all deserts are hot. The cold winds that blow across the Gobi desert of central Asia produce freezing conditions, but it is still considered a desert because it has very little rain.

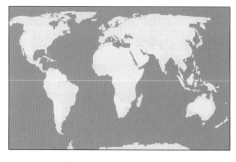

DESERT SITES
Deserts are found in subtropical areas, on the dry side of mountain ranges and in the center of large continents.

OVER THE DUNES
The strong winds in sandy deserts often create vast areas of sand dunes. The constantly shifting sand and the lack of water mean that there are few plants. Nomadic peoples often use camels as pack animals because they can survive with little water.

THE DESERT IN BLOOM
Every year, desert rains trigger the germination of thousands of plants. They grow rapidly so that they can complete their life cycle before they run out of water. A few weeks after rain has fallen, the desert is transformed into a carpet of flowers. These soon die, but their seeds lie in the ground, waiting for the next rainfall.

Creatures of the Desert

Lizards are well adapted to living in the desert. Their scaly skin enables them to retain water by cutting down evaporation. They also produce solid waste rather than liquid urine, and they can alter the color of their skin to reflect more or less heat. Desert animals have also adapted to survive. The large ears of the fennec fox enable it to lose heat from its body. The fox also has exceptional hearing and this allows it to hunt at night when temperatures are cooler. After the rains, honey pot ants collect as much nectar as they can find and give it to special ants that store the nectar in their abdomens. The ants' abdomens swell up, often to the size of grapes. During the dry months, the other ants in the colony feed off this collected honey.

Fennec fox

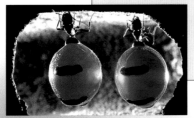
Honey pot ants

Q: Why do plants grow rapidly in the desert?

TREE RINGS
Each year, a tree grows a ring of woody material that is called an "annual ring." In warm, wet years, growth is good and the ring is wide. During years of bad weather, the ring is very narrow.

580,000 BC
Günz Ice Age

PAST ICE AGES
In the last million years, there have been four ice ages. The average temperature of the Earth during these times was 6–8 degrees below today's averages.

A MAMMOTH EVENT
Until the Würm Ice Age, huge, elephant-like mammoths lived on the cold plains that now form Siberia. The mammoths and other giant mammals disappeared as the climate became warmer and the ice retreated. Because the ice preserved their bodies, we know that they had thick, woolly coats and huge tusks.

430,000 BC
Mindel Ice Age

UNDER ICE
About two million years ago, the Earth's climate cooled and the polar ice cap expanded to cover northern Europe.

• CLIMATIC CHANGE •

Global Freezing

240,000 BC
Riss Ice Age

The Earth's climate has changed many times in the last few million years. There were periods of severe cold, known as ice ages or glacials, when great slabs of ice inched across the land. They gouged out hollows in their path as they pushed soil and rocks ahead of them. The sea level dropped enormously as much of the water froze. Warmer times between the ice ages were called interglacials. The ice melted and the huge hollows filled with water and became lakes. Scientists learn about the different climates in the Earth's history by looking for clues in nature. Some trees have lived for thousands of years and show signs of climatic changes. Fossils also provide valuable clues about wildlife and their environment. Most evidence about past climates comes by studying sediment samples from the beds of the oceans or ice samples taken from Greenland or Antarctica.

120,000 BC
Würm Ice Age

CARVED BY ICE
A glacier is a mass of ice that flows very slowly down a mountain valley. It rubs away the sides of the valley until the valley becomes a U shape.

HIDDEN CLUES
Geologists can break open rocks to find the fossilized imprint of plants millions of years after the sediment originally built up.

20,000 BC
Würm Ice Age ends

AD 1430
The Little Ice Age

The Little Ice Age

Between 1430 and 1850, northern Europe experienced a "little ice age." It was not as severe as a full ice age, but the climate became colder, crops failed and there was widespread starvation. England experienced some of the coldest winters on record during the 1810s and 1820s. The River Thames froze over regularly and Frost Fairs, where people played games and danced, were held on its icy surface. Sometimes the weather warmed without warning. People had to flee quickly as the ice beneath them began to thaw and crack.

Discover more in Polar Zones

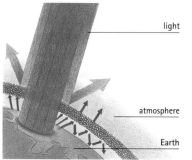

THE GREENHOUSE EFFECT
The Earth's atmosphere is like a greenhouse. It allows light from the sun to pass through it and heat the Earth's surface. Gases in the atmosphere, such as carbon dioxide, absorb the returning heat and also warm the surface. Without this greenhouse effect, the Earth would be too cold for life. But the level of carbon dioxide and other gases in the atmosphere is increasing. As more heat is absorbed by the atmosphere, the Earth becomes warmer and warmer.

• CLIMATIC CHANGE •

Global Warming

The Earth is getting warmer. Most of the hottest years during the twentieth century have occurred in the last decade. Scientists are still arguing whether this is due to the greenhouse effect or some other cause. One of the key factors in the greenhouse effect is carbon dioxide. This greenhouse gas traps heat in the Earth's atmosphere. Each year, more than 5,500 billion tons (5,100 billion tonnes) of carbon dioxide are absorbed by green plants to make food in a process called photosynthesis. This process produces oxygen, which living organisms need to breathe. However, the level of carbon dioxide in the atmosphere is increasing dramatically because of pollution, deforestation, farming methods and the burning of more fossil fuels (coal, gas and oil). Some scientists believe that the Earth's temperature will continue to rise as more carbon dioxide is released into the atmosphere.

A CHANGING WORLD
More fossil fuels are being used each year. They provide power for cars and industry, and heat homes and offices. When fossil fuels are burned, carbon dioxide is released. Cows also have a major effect on the atmosphere because they produce methane gas when they digest grass. As the number of cows increases, so does the amount of methane released.

POTENT METHANE
The greenhouse gas methane is 20 times stronger than carbon dioxide. Much of it comes from bacteria that live in waterlogged soils, such as rice paddy fields and wetlands.

A Hole in the Sky

The ozone layer, which is in the upper atmosphere, shields the surface of the Earth from ultraviolet light. But scientists have discovered that the ozone layer is being attacked by manufactured chemicals called chlorofluorocarbons (CFCs), which are used in spray cans, refrigeration and air-conditioning. The blue at the center of this picture shows a hole in the ozone layer over Antarctica. This allows more ultraviolet light to reach Australia and New Zealand. In 1987, many nations around the world signed a treaty to limit the production of CFCs. This has been effective, but the damaging effects of CFCs will last for decades.

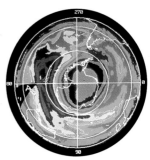

DISAPPEARING FORESTS
Trees absorb carbon dioxide and release oxygen. But forests are being cleared throughout the world. This deforestation is contributing to the increased levels of carbon dioxide in the atmosphere.

NEW LIFE
Replanting trees (reforestation) helps to reduce the greenhouse effect and fight global warming.

Discover more in Weather Watch

Beyond Planet Earth

- What is a red giant?

- Which planet is like a boiling cauldron?

- Why does the tail of a comet trail through space?

- How do the planets stay in orbit around the Sun?

AROUND AND AROUND

The planets closest to the Sun move around their orbits faster than those farther away. The Earth takes a year to complete its orbit. The planets also spin around as they orbit the Sun. The Earth spins once every 24 hours, which we call a day. Here we show the different sizes of the planets, as well as the time they take to orbit the Sun (given as a year) and to spin on their axis (given as a day).

Mercury
Year: 88 Earth days
Day: 59 Earth days

Venus
Year: 225 Earth days
Day: 243 Earth days

Earth
Year: 365.25 days
Day: 24 hours

Mars
Year: 1.9 Earth years
Day: 24.6 hours

Jupiter
Year: 11.9 Earth years
Day: 9.8 hours

DID YOU KNOW?
Pluto takes 248 years to circle the Sun and for most of that time it is the farthest planet from the Sun. But Pluto has a very oval-shaped orbit, and for 20 years of its total orbit, Pluto is actually closer to the Sun than its neighbor Neptune.

• OUR NEIGHBORHOOD •

The Solar System

Humans live on a small planet in a tiny part of a vast universe. This part of the universe is called our solar system, and it is dominated by a single brilliant star—the Sun. Our solar system is the Earth's neighborhood and the planets Mercury, Venus, Mars, Jupiter, Saturn, Uranus, Neptune and Pluto are the Earth's neighbors. They all have the same stars in the sky and orbit the same Sun. Scientists believe the solar system began about 5 billion years ago, perhaps when a nearby star exploded and caused a large cloud of dust and gas to collapse in on itself. The hot, central part of the cloud became the Sun, while some smaller pieces formed around it and became the planets. Other fragments became comets and asteroids (minor planets), which also orbit the Sun. The early solar system was a turbulent mix of hot gas and rocky debris. Comets and asteroids bombarded the planets and their moons, scarring them with craters that can still be seen today.

PLANET PATHS
The Sun is massive and has a strong gravity that pulls the planets towards it. The planets also have their own energy of motion, and without the pull of the Sun, which bends the planets' paths into orbits around it, they would fly off into space.

Saturn
Year: 29.5 Earth years
Day: 10.2 hours

Uranus
Year: 84 Earth years
Day: 17.9 hours

Neptune
Year: 165 Earth years
Day: 19.2 hours

Pluto
Year: 248 Earth years
Day: 6.4 Earth days

• OUR NEIGHBORHOOD •
The Sun

The Sun is the center of the solar system. This enormous star gives us all the light and heat we need to grow food and keep warm. It was worshipped as the mightiest of the gods by the ancient Egyptians. The Sun, however, is not the largest star in the galaxy. It seems very big and bright because it is only 93 million miles (150 million km) away from Earth. Light from the Sun takes eight minutes to reach us; light from Sirius, the next brightest star, takes eight years! The Sun is made up of gases, mostly hydrogen, and is powered by a natural process called nuclear fusion—when atoms of hydrogen fuse, or join together, to make helium. Nuclear fusion takes place in the center, or core, of the Sun, where temperatures are around 27 million°F (15 million°C). The Sun has shone in the sky for nearly 5 billion years and scientists believe it has enough hydrogen in its core to "burn" for another 5 billion years. Then it will expand to become a red giant before shrinking to become a feeble white star.

DID YOU KNOW?
The corona is as hot as the center of the Sun. However, the gases inside the corona are very thin because the gas particles are very far apart. This means that if you put your hand into this searing heat, you would not feel a thing.

CLOUD ACTIVITY
Clouds of gas called prominences can erupt from the Sun's surface. They are best seen during a total eclipse of the Sun—when the Moon cuts off the bright light of the photosphere.

INSIDE THE SUN
Energy is produced in the core of the Sun. It is transferred to the surface through the body of the star—the zone of radiation and convection. We can see the Sun's photosphere through the thin chromosphere and the outer atmosphere—the corona.

THE SURFACE OF THE SUN
This picture shows the boiling surface of the Sun. Cool, dark patches called sunspots lie beneath the bright spots seen here.

ENERGY BURST
Solar flares are huge eruptions that occur near sunspots. They release a massive amount of energy into space.

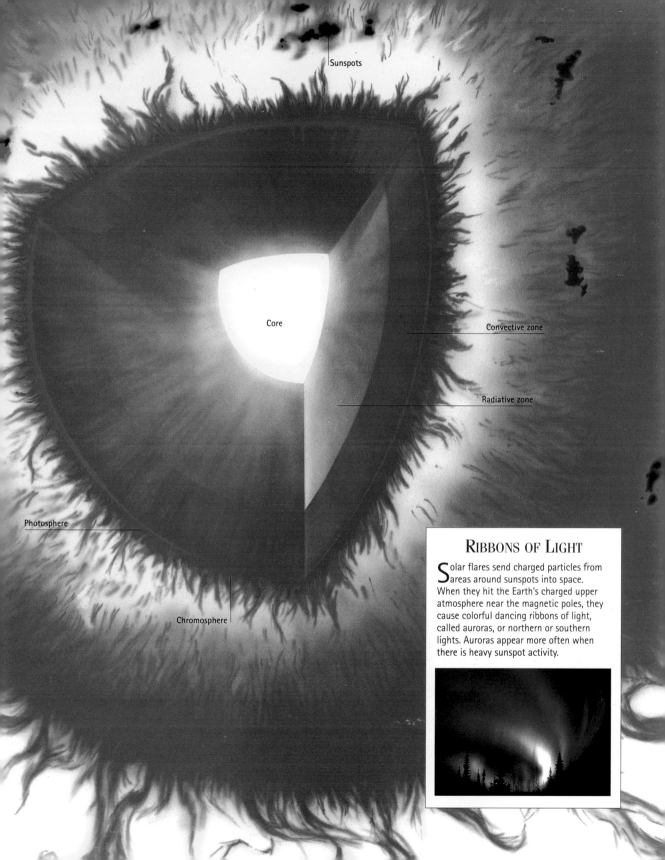

RIBBONS OF LIGHT

Solar flares send charged particles from areas around sunspots into space. When they hit the Earth's charged upper atmosphere near the magnetic poles, they cause colorful dancing ribbons of light, called auroras, or northern or southern lights. Auroras appear more often when there is heavy sunspot activity.

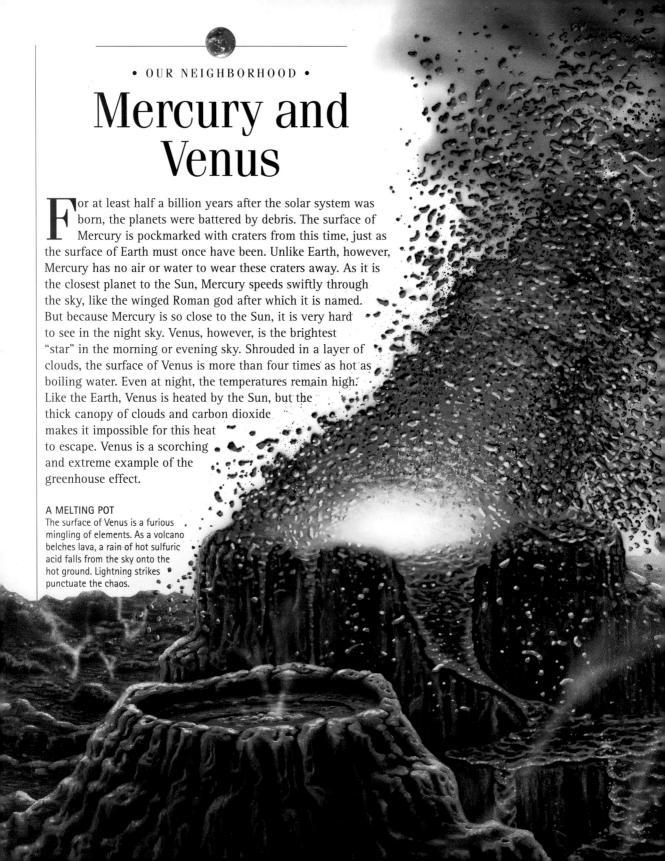

• OUR NEIGHBORHOOD •

Mercury and Venus

For at least half a billion years after the solar system was born, the planets were battered by debris. The surface of Mercury is pockmarked with craters from this time, just as the surface of Earth must once have been. Unlike Earth, however, Mercury has no air or water to wear these craters away. As it is the closest planet to the Sun, Mercury speeds swiftly through the sky, like the winged Roman god after which it is named. But because Mercury is so close to the Sun, it is very hard to see in the night sky. Venus, however, is the brightest "star" in the morning or evening sky. Shrouded in a layer of clouds, the surface of Venus is more than four times as hot as boiling water. Even at night, the temperatures remain high. Like the Earth, Venus is heated by the Sun, but the thick canopy of clouds and carbon dioxide makes it impossible for this heat to escape. Venus is a scorching and extreme example of the greenhouse effect.

A MELTING POT
The surface of Venus is a furious mingling of elements. As a volcano belches lava, a rain of hot sulfuric acid falls from the sky onto the hot ground. Lightning strikes punctuate the chaos.

Mercury, the Fossil Planet

Steep cliffs and craters scar the surface of Mercury. The day on Mercury is searingly hot because the planet is so close to the Sun, but the night is unbearably cold. As Mercury has no real atmosphere, impact craters that were formed nearly 4 billion years ago still dominate its ancient surface. In 1977, the *Mariner 10* spacecraft visited Mercury and took some revealing pictures of the planet, such as the one shown here. These photos gave us a bleak picture of the early history of the solar system. Comets and asteroids regularly hit Mercury and all the planets during a time we call the "age of heavy bombardment."

Q: Why are craters visible on Mercury but not on Earth?

Did You Know?

Venus is the Roman goddess of love, and most of the features on the planet Venus are named after real or imaginary women. Two of its continents take the names of the goddesses Ishtar and Aphrodite, and a crater is named after the famous jazz singer Billie Holiday.

A RARE EVENT
If you look very closely at this time-lapse photograph, you will be able to see a small dot, which is Venus passing behind the Moon. This rare event is called an occultation and it takes place when the Moon passes in front of a planet.

BRIGHT LIGHT
As its thick clouds reflect light back into space, Venus is by far the brightest planet in the sky. However, it is visible only before dawn or after dusk for a few months each year.

Discover more in The Solar System

179

• OUR NEIGHBORHOOD •
The Earth

We live on a small planet—the only place in the solar system where life seems to flourish. Seen from an Apollo spacecraft orbiting the Moon, the Earth is a colorful planet of green spaces, deserts, deep oceans and fields of ice. Life on Earth is possible because our planet is just the right distance from the Sun for water to exist as a liquid. If the Earth was a few million miles either closer to or farther from the Sun, it might be a boiling cauldron such as Venus, or a frozen wasteland such as Mars or the moons of Jupiter. Life is also sustained by the Earth's atmosphere—a thin layer of gas that surrounds the planet. Of all the planets in our solar system, this atmosphere is unique because it contains so much oxygen. The Earth orbits the Sun, and spins like a top once a day. This rotation and the Earth's atmosphere keep temperatures from reaching extremes, such as those on the nearby Moon.

Crust | Outer core
Mantle | Inner core

INSIDE THE EARTH
The solid iron inner core of the Earth is surrounded by a liquid outer core, and a soft rock mantle. The rock structures we actually see are part of the crust, which ranges in thickness from 3–43 miles (5–70 km).

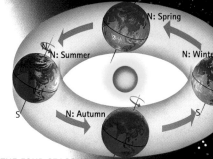

THE FOUR SEASONS
The Earth's seasons are caused by the way the Earth tilts as it orbits the Sun. Through the year, the Southern and the Northern (N) hemispheres have opposite seasons: while the Northern Hemisphere has winter the Southern Hemisphere has summer.

THE MIDNIGHT SUN
At the equator, summer days are the same length as winter days. As you go farther north or south, the difference in length between winter and summer days becomes greater and greater. If you go far enough north or south, you will reach a place (such as Norway shown below) where on some summer days the Sun never sets.

STAR TRAILS

The Sun moves across the sky as the Earth rotates. This picture shows that the stars also appear to move to the west, but it is really because the Earth is rotating towards the east. To take a picture such as this, leave your camera well-mounted and its lens open for about half an hour. As the Earth moves east, the stars will appear to draw lines on the film.

VIEWING THE MOON
At any time, half of the sphere of the Moon is lit by the Sun. We might see just a little of the lit side (at crescent phase), most of it (in gibbous, or more than half, phase) or all of it (at full phase). How much we see depends on where the Moon is in orbit around the Earth.

HIGH AND LOW TIDES
Some places on Earth, such as the Fijian coast shown here, have extreme tidal ranges. The low tides expose much of the sea floor, while the high tides seem to sweep away the land.

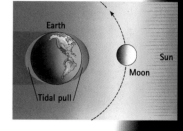

GRAVITATIONAL PULL
Tides are caused by the pull of the Moon's gravity, and to a lesser extent the Sun's gravity, on the Earth's oceans.

• OUR NEIGHBORHOOD •

The Earth's Moon

People have been entranced by the Moon for centuries. The astronomer Galileo first looked through a telescope at this mysterious ball of rock in 1609. He saw its strangely uneven surface; its mountains and craters; and its dark, lava-filled basins (called "seas"), caused by collisions that rocked the Moon during the chaotic beginnings of the solar system. From the Earth, these dark markings seem to form a pattern, which people sometimes call a rabbit, a cat or even the "Man in the Moon." Astronaut Neil Armstrong became the first man on the Moon in 1969. The world watched in awe as he stepped on this airless, waterless satellite of the Earth. As the Moon moves around its endless orbit of the Earth, it seems to change shape in our sky, depending on how much of it is lit by the Sun. But we always see the same face of the Moon because it spins on its axis in the same time it takes to orbit the Earth.

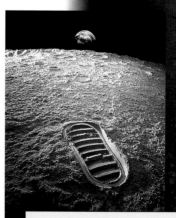

STRANGE BUT TRUE
When Neil Armstrong took "one small step for a man, one giant leap for mankind," the footprint he left was a permanent one. As there is no air on the Moon, Armstrong's footprint should last for many millions of years. Eventually, tiny hits from small meteoroids will cause the footprint to fade.

IN THE BEGINNING

How the Moon came to be is a subject of great debate. The best of the current theories says that a tremendous collision, early in the Earth's history, produced a cloud of rocky debris that orbited the Earth. The debris formed clumps that heated as they collected together. The result was a new body that cooled down to become the Moon.

• OUR NEIGHBORHOOD •

Mars

The Romans called the orange-red planet in the night sky Mars, after the god of war. Its surface is covered by rusty-red rock and dotted with huge canyons and volcanoes, polar icecaps and mountains. Phobos and Deimos, tiny moons scarred by craters, orbit the planet. Since the astronomer Schiaparelli first studied Mars in the late nineteenth century, people have wanted to believe that there was life on this planet. It is only half the size of Earth, but the planets are similar in some ways: the day on Mars is half an hour longer than ours, and it has changes in weather like our seasons. In the 1970s, space probes visited Mars, but their findings showed that the red, rocky planet is like a chillingly cold desert. Water probably lies frozen beneath the hostile ground. The atmosphere on Mars is too thin to breathe, and violent dust storms sometimes howl across its surface.

THE LARGEST OF THEM ALL
Olympus Mons is the biggest volcano in the solar system. It is as large as the American state of Arizona! Olympus Mons rises so slowly that you could climb it without being aware that you were getting higher.

INTO THE FUTURE
A spacemobile such as this may one day be used to carry scientists across the surface of Mars to collect specimens from canyons and ancient river beds.

MARS ROVER
This vehicle has been built especially for exploring the surface of Mars. Its large wheels will help it travel across the rough terrain.

LIFE ON MARS

About a century ago, an Italian astronomer named Schiaparelli sketched Mars. On nights when the sky was especially clear, he could see long lines, which he called channels, crisscrossing the planet. American astronomer Percival Lowell (above) became very interested in Schiaparelli's observations. He developed a theory that these lines were really canals, built by a Martian civilization to transport the planet's dwindling water supply. Many people wanted to believe him, but in 1965 *Mariner 4* passed near the planet and saw no sign of any such constructions.

DID YOU KNOW?

On Halloween night in 1938, many people in the United States tuned in to their radios as actor Orson Welles narrated the dramatic science-fiction story *The War of the Worlds* by H. G. Wells. Before the program was over, millions of people believed that Martians were really invading the Earth.

A FAMILY OF MOONS
Jupiter is surrounded by a large family of moons. Galileo saw Io, Europa, Ganymede and Callisto, the biggest and brightest of all Jupiter's moons, through a telescope in 1610.

COLLISIONS IN THE UNIVERSE
In July 1992, Comet Shoemaker–Levy 9 passed so close to Jupiter that it split into 21 pieces. Two years later, the comet fragments collided with Jupiter. Every large telescope on Earth and in space was poised to see the dramatic collision and the huge, spectacular fireballs that rose about 1,900 miles (3,000 km) above Jupiter's clouds.

• OUR NEIGHBORHOOD •

Jupiter

Named after the king of the Roman gods, Jupiter is the largest planet in the solar system. It is 300 times heavier than the Earth and more than twice as heavy as all the other planets added together! The enormous gravity causes very high temperatures and pressure deep inside Jupiter. This stormy planet is cloaked by noxious gases such as hydrogen, ammonia and methane and topped by bitterly cold, swirling cloud zones, which change in appearance as the planet spins quickly on its axis. A day on Jupiter is less than 10 hours long—the shortest day in the solar system. This speedy rotation causes great winds and wild storms. Like most of the planets, Jupiter has moons. We know of at least 16, but there may be smaller moons still to be found. In 1979, the *Voyager 1* space probe discovered that Jupiter was encircled by a narrow, faint ring made up of rocky or icy fragments.

THE MOONS OF JUPITER

Jupiter's four largest moons are very different from each other. Io, the closest of the four to Jupiter, has many volcanic vents that spew clouds of sulfur into the sky. *Voyager 1* recorded five erupting volcanoes when it visited Io in 1979. The smooth, icy surface of the next moon, Europa, is patterned with cracks that may once have been filled with water. The icy surfaces of Ganymede and Callisto, the largest moons, are marked by craters, the remains of ancient impacts.

Io

Europa

AS STORMY AS EVER
Astronomers first saw Jupiter's Great Red Spot 300 years ago. This swirling whirlpool of gases is a huge storm cloud that seems to rage constantly on this turbulent planet.

Q: Why is Jupiter such a stormy planet?

• OUR NEIGHBORHOOD •

Saturn

> **DID YOU KNOW?**
> Early astronomers could not see Saturn's rings clearly. However, they are seen easily through today's telescopes, except when Saturn is tilted at a particular angle and the rings are side-on to Earth.

The bright rings of Saturn are a dazzling highlight of the night sky. Ever since Dutch scientist Christiaan Huygens saw them through a telescope more than 300 years ago, astronomers have turned their sights to Saturn. From the Earth, it seems that Saturn is surrounded by three rings, but the 1981 Voyager space probe discovered that there are thousands of narrow ringlets made up of millions of icy particles. These ringlets stretch for thousands of miles into space in a paper-thin disk. The rings were formed long ago, perhaps when a moon or an asteroid came too close to Saturn and was torn apart by the strong gravity of the planet, which is the second largest in the solar system. Like Jupiter, Uranus and Neptune, Saturn is made up mainly of hydrogen and helium. It spins very quickly on its axis and is circled by bands of clouds.

Cassini's division
This has far fewer ringlets.

A LASTING IMPACT
One of Saturn's moons, Mimas, has a huge crater called Herschel. It was caused by a violent collision with a comet or an asteroid long ago, which nearly tore the little moon apart.

A ring
This is very bright where the ringlets are close together.

B ring
The color of this seems to be more solid.

STUDYING SATURN

Saturn has a smooth, yellowish tinge, which is caused by a layer of haze that surrounds the planet. Unlike Jupiter, which it resembles slightly in color, it does not seem to have any longlasting light or dark spots.

C ring
From Earth, this is seen as a faint ring.

Encke gap
This is a large gap within the A ring.

F ring
Seems to be knotted or braided.

Q: What are the names of the other gas planets?

SATURN'S MOONS

Saturn has more moons than any other planet. Eighteen have been discovered so far, but there are probably smaller moons that we have not seen yet.

FAR FROM THE SUN

Orange-colored Titan is Saturn's biggest moon. It is the only moon in the solar system with an atmosphere, which is made up mostly of nitrogen. As on Venus, this thick atmosphere hangs over Titan like a veil. Because it is so far away from the Sun, Titan is very cold and the methane on the planet is a liquid, not a gas.

• OUR NEIGHBORHOOD •

Uranus

Uranus is the Greek god of the sky. The planet was first noticed in 1781 by Englishman William Herschel, who saw a small round object with a greenish tinge through his home-made telescope. The discovery of Uranus caused great excitement. Astronomers had previously believed that Saturn lay at the edge of the solar system. As Uranus lies twice as far from the Sun as Saturn, the known size of the solar system suddenly doubled! Uranus is nearly four times the size of the Earth, and it orbits the Sun every 84 years. Like Jupiter and Saturn, it is made up mainly of hydrogen and helium. Most of the planets in the solar system are tilted a little (the Earth is tilted at an angle of 23°, and this causes the different seasons), but Uranus is tilted completely on its side. This means that each pole has constant sunlight for 42 years. When the *Voyager 2* space probe passed Uranus in 1986, it photographed the dense clouds that cover the planet, its narrow rings and its beautiful moons.

THE MOON MIRANDA
Miranda, the smallest and most unusual of the five main moons of Uranus, looms into view in front of the planet. This photograph was taken by the *Voyager 2* space probe as it flew past Uranus in 1986.

CLOSE-UP
Miranda has a unique surface. Some astronomers believe that it may have broken apart after a collision, but then re-formed. When the pieces of the moon came back together again, its surface was buckled with deep grooves.

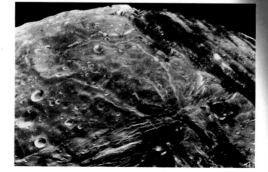

Q: Why was the discovery of Uranus so exciting?

BEFORE
In the early days of the solar system, a large body may have crashed into Uranus.

AFTER
This collision tilted Uranus so that it rolls through the sky on its side.

DID YOU KNOW?

The rings of Uranus were discovered by astronomers in 1977. As Uranus passed in front of a star, they saw that the star's light flickered. They realized that rings around the planet were blocking the light of the star as they passed across it.

THE ROYAL ASTRONOMER

When William Herschel first observed Uranus, he thought he had discovered a comet or a star. His find proved to be much greater. King George III of England was so delighted with the discovery, he made Herschel his private astronomer. The king also gave Herschel the funds he needed to build larger telescopes, such as this.

Discover more in The Solar System

• OUR NEIGHBORHOOD •

Neptune

Neptune is the smallest of the four gas planets and more than 2 billion miles (3 billion km) away from the Sun. Astronomers see a faint star when they view Neptune through a small telescope. This deep-blue planet is a bleak and windy place, with poisonous clouds made of methane ice crystals swirling around it. The planet's rocky core is about the size of the Earth, and is surrounded by a frozen layer of water and ammonia. Like the other gas planets, Neptune's atmosphere consists mainly of hydrogen. Neptune was discovered in 1846, but until *Voyager 2* sent pictures of it back to Earth in 1989, we understood very little about it. Now we know that Neptune has many faint rings and eight moons. The largest moon, Triton, is covered by ice and has mysterious features such as dark streaks. These could be caused by volcanoes erupting nitrogen, which becomes a liquid in Triton's intensely cold climate.

A BRIEF SPOTTING
Voyager 2 photographed this spinning storm cloud called the Great Dark Spot. Five years later, however, photographs from the Hubble Space Telescope showed that the spot had disappeared.

RULER OF THE SEVEN SEAS
The Romans believed that Neptune was the powerful god of the sea. His son Triton, who was half man and half fish, ruled the stormy waves with his father.

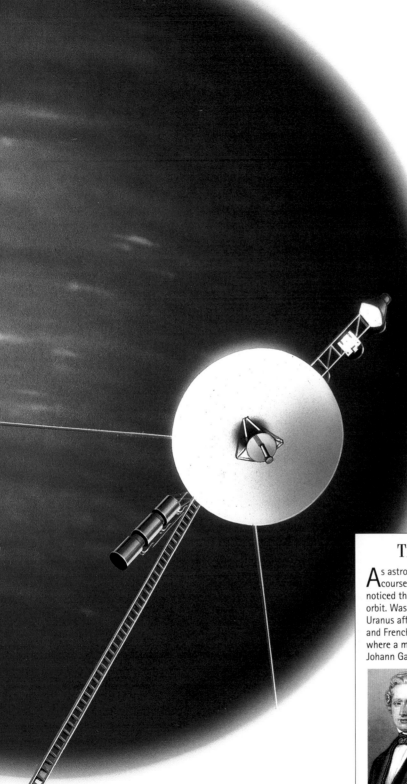

Q: How many moons does Neptune have?

SURFACE ERUPTIONS
Triton, the Earth, Venus and Jupiter's moon Io are the only places in the solar system where there seems to be volcanic activity. But Triton's volcanoes are cold and erupt liquid nitrogen, not hot lava.

PICTURING NEPTUNE
The American space probe *Voyager 2*, shown here, was launched in 1977. Twelve years later, it reached Neptune on the distant edge of the solar system. Radio messages from the probe traveled to Earth at the speed of light.

The Discovery of Neptune

As astronomers in the nineteenth century plotted the course of the stars and planets in the solar system, they noticed that Uranus did not seem to follow its predicted orbit. Was the gravity from an undiscovered planet beyond Uranus affecting its orbit? Englishman John Couch Adams and Frenchman Urbain Le Verrier both calculated exactly where a mystery planet might lie. The German astronomer Johann Galle used their careful research, and in 1846 he became the first person to see the new planet through his telescope.

Far left: Urbain Le Verrier
Left: John Couch Adams

• OUR NEIGHBORHOOD •

Pluto

Pluto lies in the far reaches of the solar system and is named after the Greek god of the dark underworld. This rocky planet is the smallest in the solar system and, usually, the farthest from the Sun. Pluto, however, has a strangely elongated orbit. It spends 20 years of the 248 years it takes to orbit the Sun inside the orbit of Neptune. Then it moves away and heads deeper into space. For most of Pluto's long year, the materials that make up its surface are frozen. But when Pluto moves closer to the Sun, some of these materials turn from solids into gases, and the planet has an atmosphere. Pluto was not discovered until 1930, and it has not yet been reached by a space probe. Although we know that it has a moon, called Charon, which takes about six days to circle Pluto, there is still much to learn about this distant speck in the night sky.

THE VIEW FROM SPACE
Pluto and Charon are very close together. They loom largely in each other's night sky.

Q: Why is Pluto's orbit so unusual?

DID YOU KNOW?

Walt Disney, the famous American film-maker, created the droopy-eared cartoon character Pluto just a few months after the planet Pluto was found and named.

LOOPING THE LOOP
This diagram shows the oval-shaped orbit of Pluto around the Sun. Twice during its 248-year orbit, Pluto's path brings it closer than Neptune to the Sun.

THE VIEW FROM PLUTO
The shadow of the moon Charon falls on the icy surface of Pluto. Charon is half the size of Pluto, and Pluto is smaller than the Earth's moon.

THE SEARCH FOR THE MYSTERY PLANET

Astronomer Clyde Tombaugh (right) discovered Pluto in 1930. He had followed Percival Lowell's theories that a planet lay beyond Neptune, which could explain the irregular orbits of Neptune and Uranus. But astronomers soon realized that the discovery of Pluto did not explain this at all! Pluto was far too small to have such an effect on the planets' orbits. Some astronomers think that there is another, more massive planet farther away from Pluto. The search continues.

• OUR NEIGHBORHOOD •

Comets

Comets are icy balls that sweep through the solar system. Long ago people thought these "long-haired stars," which appeared mysteriously and dramatically in the sky, were a sign that evil events were about to happen. Edmund Halley dispelled this idea in the eighteenth century by proving that comets, like all matter in the solar system, have set orbits around the Sun. Some comets pass near the Sun every few years. Others have long orbits and pass close to the Sun only once. As a comet gets closer to the Sun, its nucleus (center) begins to warm up and gives off a cloud of dust and gas called a coma. Astronomers can see the coma through a telescope because it reflects the fiery light of the Sun and becomes much larger than the Earth. As the comet journeys towards the Sun, the solar wind blows a stream of dust and gas away from the comet and the Sun. This forms the comet's tail, a spectacular streak of gas and dust that can trail for millions of miles into space.

STRANGE BUT TRUE
People in the past believed that comets brought disasters. In the fifteenth century, astrologers for the Archduke of Milan told him he had nothing to fear, except a comet. Unfortunately, a comet appeared in 1402. The archduke was seized with panic when he saw it, had some kind of attack, and died.

Gas tail
This is straight, narrow and usually fainter than the dust tail.

Coma
This envelope of gas surrounds the nucleus of the comet.

Nucleus
This is a mixture of ice and dust. It gives off a cloud of dust and gas when it is heated.

CLOSE-UP OF A COMET
The nucleus is the dirty snowball at the heart of a comet. It is so small that it cannot be seen from Earth by the naked eye. But we can see the huge coma, and the gas and dust tail (or sometimes tails) that stream behind the comet.

COMET'S TAIL
As a comet orbits the Sun, its tail grows and fades, but always points away from the Sun.

Dust tail
This is usually curved and is made up of gases pushed away from the Sun by the solar wind.

Return of the Comet

Halley's Comet is probably the most famous of all comets. Edmund Halley (top) was the first person to calculate that the appearance of three separate comets through the years was, in fact, the return of one comet every 76 years. The comet was named after him when he successfully predicted its return in 1758. In early times, Halley's Comet terrified those who saw it. In 1910, we had an opportunity to view the comet at close range as the Earth passed through the comet's tail. In 1986, spacecraft from different nations went out to meet Halley's Comet (bottom). The comet is now beyond the orbit of Uranus, east of the constellation of Orion. In 2062, Halley's Comet will once again brighten the sky.

COMET LEVY
Comets are often named after the people who first saw them. In 1990, Comet Levy lit the sky all night long.

• OUR NEIGHBORHOOD •

Asteroids and Meteoroids

The solar system has many different members, the smallest of which are asteroids and meteoroids. Asteroids are small rocky bodies that never came together in the early days of the solar system to form larger planets. Most asteroids lie in the enormous space between the orbits of Mars and Jupiter, an area called the asteroid belt. Ceres, the largest of these asteroids and the first to be discovered, is almost 500 miles (800 km) wide. Most asteroids, however, are much smaller. Meteoroids are dust particles that travel along the orbital paths of comets. When a meteoroid encounters the Earth's upper atmosphere at high speed, it usually burns up and forms a bright meteor. Some people call this brief streak of light a "shooting star." Larger meteors that pierce the Earth's atmosphere and crash to the ground, making craters where they land, are called meteorites.

ASTEROID ORBITS

Not all asteroids orbit in the main belt between Mars and Jupiter (the larger planet in this picture). Two groups of asteroids called Trojans share Jupiter's orbit. Other asteroids cross the orbit of the Earth.

FLYING OBJECTS IN SPACE

On its June 1983 voyage, the space shuttle *Challenger* was hit by a tiny particle, possibly a meteoroid or a speck of paint left by a spacecraft on a previous mission. The shuttle and the small particle were traveling so fast that the impact left a small crater in the shuttle's window (below). Even the tiniest object moving at high speed is dangerous in space.

DID YOU KNOW?

In 1992, an asteroid called Toutatis passed close to Earth and astronomer Steve Ostro was able to bounce radar signals off it. From the reflection on the radar, he was able to discover that Toutatis actually looked like two asteroids close together.

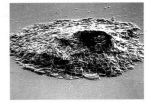

LOOK OUT BELOW!

These children are standing next to one of the largest meteorites in the world. Long ago, the Inuit (Eskimos) of Cape York in Greenland worshipped this object from the sky.

MAKING ITS MARK

Astronomers calculate that large asteroids collide with the Earth every hundred thousand years or so. This crater at Gosse Bluff in the Northern Territory of Australia is the result of such a collision.

Discover more in The Solar System

Q: Where are most asteroids found?

STAGES OF A LUNAR ECLIPSE
These photographs show the progress of a lunar eclipse, from the first bite of the Earth's shadow on the Moon's surface to the total phase and beyond.

• OUR NEIGHBORHOOD •

Eclipses

The Sun sends its light far into space. As light falls on the Earth and the Moon, both cast a shadow. An eclipse of the Moon (a lunar eclipse) occurs when the shadow of the Earth darkens the Moon. The Moon sometimes becomes coppery red or even brownish as the Earth's shadow marches across its surface. An eclipse of the Sun (a solar eclipse) occurs when the Moon passes in front of the Sun and blocks the light to places along a narrow strip of the Earth's surface. A strange darkness falls on the land and temperatures drop. The Moon is 400 times smaller than the Sun, but it can hide the light of such an enormous star because the Sun is so far away from the Moon. When we look at the Sun and the Moon in the sky, they appear to be almost the same size. Solar eclipses occur in cycles. One eclipse will be very similar to another that happened more than 18 years earlier, but they will not be at the same place.

CLOAKING THE SUN
Eclipses can be either partial, if only part of the Sun or Moon is covered; or total, if the whole is hidden from view. This time-lapse photograph shows the progress of a total eclipse of the Sun. The Moon takes just over an hour to cover the Sun completely. It hides the light from the Sun's surface and allows us to see the Sun's faint, ghostly corona.

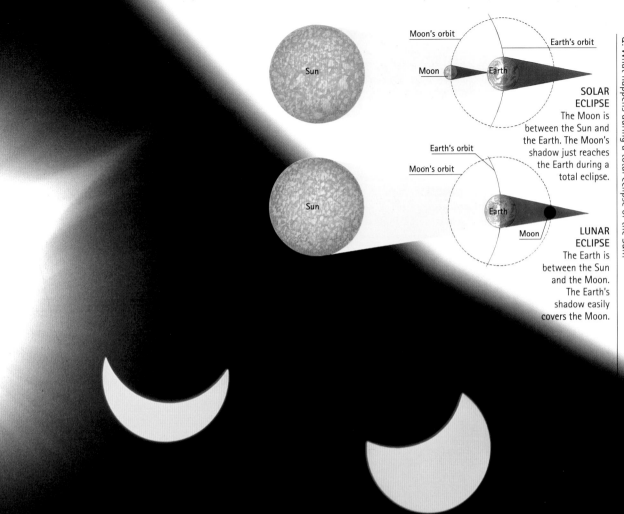

Q: What happens during a total eclipse of the Sun?

SOLAR ECLIPSE
The Moon is between the Sun and the Earth. The Moon's shadow just reaches the Earth during a total eclipse.

LUNAR ECLIPSE
The Earth is between the Sun and the Moon. The Earth's shadow easily covers the Moon.

VIEWING A SOLAR ECLIPSE

Eclipses are unforgettable sights that we would all like to see, but the Sun is very dangerous to look at without proper protection for your eyes. Permanent blindness can result from the shortest look through binoculars or telescopes. The eyepiece filters that are often supplied with small telescopes are not safe, either. The girl in this picture is safely viewing an eclipse without a telescope. With the Sun behind her, she holds a piece of paper with a hole through it. The light passes through the hole and projects an image of the eclipse onto another piece of paper in front of her.

Discover more in The Sun

• OUR UNIVERSE •

The Universe

> ### DID YOU KNOW?
> Scientists measure the vast distances in the universe in light years—the distance light travels in a year. It travels at 186,000 miles (300,000 km) per second, which is equal to seven times around the Earth every second. The Andromeda Galaxy is 9 quintillion miles (14 quintillion km) from here. It is impossible to imagine this huge distance, so we say that it is 2 million light years away.

The Earth and the other planets, the stars, the galaxies, the space around them and the energy that comes from them are all part of what we call the universe. Most astronomers believe that between 8 and 16 billion years ago, all matter and energy, even space itself, were concentrated in a single point. There was a tremendous explosion—the Big Bang—and within a few minutes the basic materials of the universe, such as hydrogen and helium, came to be. These gases collected together into large bodies called galaxies. Today, the universe still seems to be expanding. Huge families, or superclusters, of galaxies are racing away from all the other clusters at incredible speeds. If the Big Bang has given them enough energy, the galaxy superclusters may keep on racing away from each other until the last star has died. But if their gravity is strong enough to slow them down, everything in the universe will eventually cascade in on itself in an event we call the Big Crunch. Then, perhaps another cycle will begin.

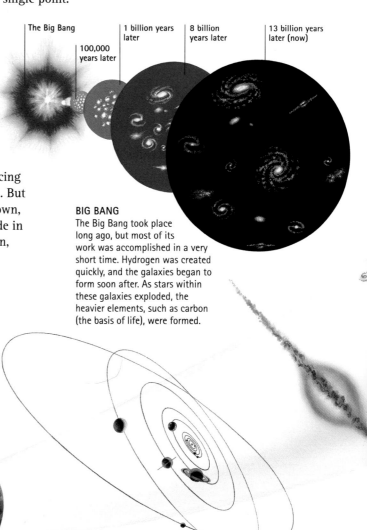

The Big Bang | 100,000 years later | 1 billion years later | 8 billion years later | 13 billion years later (now)

BIG BANG
The Big Bang took place long ago, but most of its work was accomplished in a very short time. Hydrogen was created quickly, and the galaxies began to form soon after. As stars within these galaxies exploded, the heavier elements, such as carbon (the basis of life), were formed.

A SMALL PART OF A LARGE SCHEME
We live on the Earth, just one planet in the solar system. Our solar system is part of the Milky Way, just one galaxy in a cluster of galaxies. These clusters gather into superclusters of galaxies, all of which are expanding outward.

NIGHT SKY
The ceiling of stars we can see on a clear night is a tiny part of the universe, which is immense in both time and space.

BIG CRUNCH
The expansion of the universe will be reversed if gravity is strong enough to pull everything together again.

Close   Closer   Closer still   The Big Crunch

BACK TO THE PAST

In 1965, scientists Arno Penzias and Robert Wilson were testing a radio antenna when they detected strange energy emissions. They searched for the source of these emissions and soon made a staggering discovery: the universe had a very weak level of radiation. The existence of radiation confirmed the theory of some astronomers that the Big Bang had left a cool afterglow in space. In 1978, Penzias and Wilson won the Nobel Prize in Physics for discovering this important fact about the beginning of the universe.

Discover more in Galaxies

• OUR UNIVERSE •

The Life Cycle of a Star

Nebula
This is a huge cloud of hydrogen, helium and microscopic dust.

Picture a huge, dark cloud (a nebula) in space. When a nearby star explodes, a shock wave travels through the cloud. The cloud begins to shrink and divide into even smaller swirling clouds. The center, called the protostar, gets hotter and hotter until it ignites and a new star is born. All the stars in the sky were born from clouds of gas and dust. The hottest stars are blue-white in color and burn their hydrogen fuel very quickly. The Sun, a small yellow star, burns hydrogen more steadily. Proxima Centauri, the closest star to the Sun, burns its gas very slowly and is a cool, red star. The speed at which the stars burn hydrogen determines how long they will live. Blue giants have a short life, and explode dramatically. The Sun will continue to burn for another 5 billion years. Then it will expand into a large red giant and finally shrink to a white dwarf. Proxima Centauri, however, will remain unchanged for tens of billions of years.

THE ORION NEBULA
Orion is one of the best known groups of stars (constellations) in the sky. Some 1,600 light years away and 25 light years wide, the Great Nebula in Orion is a stellar nursery, a place where new stars are being born out of interstellar gas.

White dwarf
After the planetary nebula disappears, all that remains is a small, hot, faint star.

THE CYCLE OF LIFE
The main diagram shows the different stages in the life of a star such as the Sun. This kind of star lasts for many billions of years. As it uses up its hydrogen, it begins to swell and will become, briefly, as large as the orbit of the Earth. Then it will shrink to become a white dwarf, slowly cooling for many billions of years.

Planetary nebula
Later in its life, a star slowly blows off its outer layers to form a planetary nebula that eventually disappears.

Protostar
The center of the nebula gets hotter as it shrinks, finally creating a new star.

Life of a star
Massive stars live for perhaps several hundred million years. Smaller stars last for many billions of years.

Red giant
Late in its life, a star grows to form a red giant with an enormous surface area.

A BLACK HOLE

After a very heavy star uses up its hydrogen and explodes as a supernova, its core becomes smaller and smaller until finally it is smaller than the head of a pin. The star, however, still has gravity. This is so strong that even light from a few miles around the star cannot escape. This is called a black hole.

A SUPERNOVA
For a few days, a supernova (above left) can outshine an entire galaxy of hundreds of billions of stars. Then it becomes a tiny dot, as shown by the arrow.

THE HORSEHEAD NEBULA
A nebula is a bright or dark cloud made up of gas or dust, or sometimes both. The Horsehead Nebula is very dark and can only be seen against a background of stars or a bright nebula.

Q: What is a red giant?

Discover more in Asteroids and Meteoroids

• OUR UNIVERSE •

Strange Stars

> **DID YOU KNOW?**
> In 1931, Clyde Tombaugh discovered a variable star while searching for new planets. Most of the time, Tombaugh's star, TV Corvi, is invisible. However, it explodes at regular intervals, after which it can be seen for a brief period.

Stars are giant balls of hot gas. Their range of size, color, temperature and brightness varies enormously. They can be members of a pair, triplet or a huge cluster of hundreds or thousands of stars. The color of a star indicates how hot it is: cool stars are red, hot stars are bluish. Many of the stars studied by astronomers are in pairs and orbit each other. These "binary stars" often differ in brightness and color: a dim white dwarf, for example, might orbit a red giant. Stars that make up a binary pair are usually a great distance from each other, but some are so close they almost touch. These stars are called contact binaries and as they are so close, they have to orbit each other very rapidly. The smaller star is very dense and its gravity constantly sucks hydrogen gas away from the larger star. The big star becomes distorted and turns into a distinctive teardrop shape.

A CHANGE OF SIZE
Some stars grow bigger and smaller, as shown above. The most famous and important of these are called Cepheid variables. Their color, temperature and brightness change with their size.

QUICK FLASHES
A small, extremely dense neutron star is often all that remains of a star after it has become a supernova. It rotates in a second or less, and if we see the quick flashes we call it a pulsar.

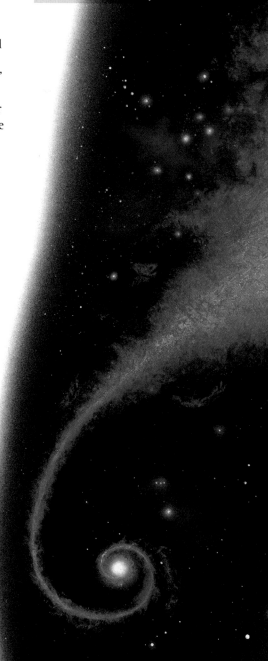

◊ MEETING OF MATTERS
In this contact binary system, the large star looks much more impressive and powerful. But the small, dense star has the real power. It has a stronger gravitational pull than the big star and is able to drag matter away and distort the shape of the large star.

EXPLODING STARS

What happens to all the hydrogen that the small star takes from the large star in a binary star system? The small star has no use for it and the hydrogen collects in a disk. When enough hydrogen has built up, over months, decades or possibly centuries, it blows up in a huge nuclear explosion. The star brightens 100 times or more for a few days. Once the explosion has died away, the process begins again.

Discover more in The Life Cycle of a Star

• OUR UNIVERSE •

Galaxies

Galaxies are enormous families of stars that lie scattered across the never-ending space of the universe. Each galaxy contains many millions of stars—a mixture of giant and dwarf stars, old and young stars, and clusters of stars. Some galaxies are spiral in shape, while others are elliptical (like a flattened circle). Those that do not seem to have much of a shape at all are called irregular galaxies. There are countless numbers of galaxies, and they are grouped together in clusters. Our solar system, for example, is part of the Milky Way Galaxy. This belongs to a collection of galaxies called the Local Group, which contains about 25 galaxies, such as the Large and Small Magellanic Clouds. The Andromeda Galaxy, the largest member of our cluster, is so huge we can see it in a very dark sky without a telescope. It lies more than 2 million light years away from Earth. Light reaching us now from the Andromeda Galaxy began its journey across space long ago when the earliest humans lived on Earth.

NEIGHBORING GALAXY
The spiral Andromeda Galaxy is the nearest major galaxy to the Milky Way. It contains hundreds of billions of stars and its spiral arms are mottled with bright and dark nebulae.

HUBBLE DEEP FIELD
This view of the universe, made by combining a series of images, was taken by the Hubble Space Telescope in December 1995. It shows several hundred galaxies never seen before.

SPIRAL
The arms of a normal spiral galaxy are filled with stars and gas clouds.

BARRED SPIRAL
A barred spiral galaxy has a bar of stars across its center. The spiral arms begin at the ends of the bar.

DID YOU KNOW?
The Magellanic Clouds are satellite galaxies of the Milky Way. They were named after Portuguese explorer Ferdinand Magellan who observed them in the early 1500s. In the future, the Milky Way's gravity may pull these galaxies apart. Their stars and nebulae will then become part of our galaxy.

ELLIPTICAL GALAXY
Giant elliptical galaxies are massive. This galaxy has 5 trillion stars.

A LIGHT FROM THE EDGE OF THE UNIVERSE

Quasars are extraordinarily powerful beacons, scattered deep in the universe. The word "quasar" stands for "quasi-stellar" (resembling a star), but quasars have far more energy than stars. A quasar called 3C-273 is several billion light years away from Earth, yet it is bright enough to be seen with a large amateur telescope. Such brilliance suggests huge size, but quasars are probably less than one light year across. Astronomers believe that a quasar is a black hole at the center of a distant galaxy, which consumes all the matter around it. The whirling matter being sucked into the hole creates an amazing source of energy and powerful "jets" of material (right), which are projected out of the galaxy's glowing core.

IRREGULAR GALAXY
Irregular galaxies have random shapes, and they are smaller than the Milky Way.

A VIEW FROM THE UNIVERSE
The Milky Way would look like a mighty spiral of stars, gas clouds and dust if viewed from one of the distant globular clusters.

• OUR UNIVERSE •

The Milky Way

On a clear night, the sky is speckled with thousands of stars. In fact, these are just a few of the 200 billion stars belonging to our galaxy, the Milky Way. From Earth, the thickest part of the Milky Way looks like a patchy band of white light stretching into space. But the Milky Way is actually shaped like a spiral, and is about 100,000 light years across. It has at least two major arms, made up of dusty nebulae and brilliant blue-white stars. Older yellow and red stars form the nucleus of the galaxy. As Earth lies way out on one of the arms of the galaxy, 30,000 light years from the center, it is very difficult for us to imagine what the galaxy looks like from the outside. Clouds of dust and gas also block much of our view of the middle of the galaxy. Astronomers, however, have recently determined that a huge object, possibly a black hole, lies in the center of the Milky Way.

DID YOU KNOW?

The Milky Way is shaped like a pinwheel, but it does not rotate as one large disk. Different stars move at different speeds. The Sun, for example, takes 220 million years to complete a trip around the center of the Milky Way.

STUDYING THE MILKY WAY

Astronomer Bart Bok (below) and his scientist wife Priscilla Fairfield devoted their lives to unraveling the mysteries of the Milky Way. By careful observation, they mapped out the spiral arms of the galaxy. They also studied the great clouds that illuminate the sky in the constellations of Orion and Carina, and tried to piece together how new stars are born from these clouds.

OUTSIDE LOOKING IN
If you could look at the Milky Way from the outside, you would see a central bulge surrounded by a thin disk that contains the spiral arms.

Discover more in Galaxies

• EXPLORING THE UNIVERSE •

Into Space

In 1957, Russia launched the first artificial satellite, *Sputnik 1*, into orbit. The Space Age had begun. Four years later, Russian Yuri Gagarin became the first person in space, and President Kennedy declared that the United States would put a man on the Moon by the end of the decade. In 1969, the *Apollo 11* spacecraft, attached to the biggest rocket ever built, pierced the Earth's atmosphere and headed towards the Moon. Astronaut Neil Armstrong's first words and steps on the scarred surface of the Moon soon became history. In the last 30 years, we have explored and discovered much about space, the stars and the planets. Spacecraft have flown past all the planets and their moons, except Pluto on the edge of the solar system; there have been several expeditions to the Moon; and space probes have landed on both Mars and Venus. Space shuttles are sent regularly into space as workhorses. Their crews sometimes repair the many satellites orbiting the Earth, such as communication satellites, which send telephone and television signals all around the world.

PROBING SPACE
The Japanese space probe *Tenma* looks at objects in space, such as black holes and supernovas, which have much energy.

LIVING IN SPACE
This is an imagined space station of the future. It is much bigger than the Soviet Mir space station, which was launched in 1986. People are aboard Mir most of the time, and the longest stay by one person so far has been a year.

DID YOU KNOW?

The American space probe *Magellan*, launched in 1989, spent four years circling Venus. Using radar, it mapped 98 per cent of the planet, revealing very clearly its volcanoes, and the mysterious channels, perhaps caused by lava flows, which slice through its surface.

LAUNCHING INTO SPACE

Space shuttles are very special spacecraft. They are designed to be used many times, taking heavy objects, such as space probes, around the Earth; and carrying crews for scientific research into space. They have three parts: an orbiter, an external tank, and two solid rocket boosters. The first space shuttle in 1981 carried two astronauts. Today, they can take a crew of up to eight people.

ON THE MOON

Even the simplest tasks require great planning and patience in the low gravity of the Moon. This astronaut is collecting a rock sample, which will be examined back on Earth.

Discover more in The Earth's Moon

• EXPLORING THE UNIVERSE •
Imagined Worlds

Is there life on other planets? Astronomers and science-fiction writers have considered this question for years. Aliens, mutant monsters and other life forms, both menacing and friendly, have starred in many books, films and series, such as *Star Wars* and *Star Trek*. People all over the world regularly report sightings of Unidentified Flying Objects (UFOs) and encounters with strange beings from space. Is this fact or fiction? Many astronomers believe that life-forming conditions do exist elsewhere in the Milky Way. For many years, people looked to our solar system and thought that Mars was the other planet most likely to support life. Technology has now made it possible to study Mars in detail and this idea today seems unlikely. Much of space, however, remains unknown territory. Science-fiction writers imagine worlds and events beyond our own. In 1865, writer Jules Verne predicted that we would reach the Moon. Some of our imagined worlds may also come true.

A CITY ON MARS
In an imagined world on Mars, a spaceship prepares to land on the red surface of the planet. A space base, where people live and carry out research, has been built in the shelter of a deep valley.

ANYONE OUT THERE?

In New South Wales, Australia, the Parkes radio telescope has been listening to the heavens for more than 30 years. But in 1995, as part of Project Phoenix (a worldwide search for life in outer space), the telescope examined areas around some nearby stars for regular signals that could come from intelligent life. The normal levels of radiation in the universe produce a random noisy hiss. If a radio telescope picks up a more orderly signal, such as that from a radio transmission, this could be evidence that life exists elsewhere.

WORLDS AWAY
The covers of these science-fiction magazines from the 1920s show strange forms of life on Neptune (right) and Venus (far right).

Strange but True

The distances in space are astronomical. In order to move from one star system to another, the makers of *Star Trek* developed the idea of traveling at what they called warp speed, which is many times faster than the speed of light. Such speed would be essential to make travel between the stars possible within a human lifetime.

Facts and Figures

SOLAR ECLIPSES
Solar eclipses can be partial (when only a part of the Sun is blocked from the Earth's view), total (when the Sun is totally blocked from the Earth's view) or annular (when the Sun's light is still visible around the edge of the Moon).

DATE	TYPE OF ECLIPSE	AREA FROM WHICH ECLIPSE CAN BEST BE VIEWED
April 17, 1996	partial	New Zealand, Antarctica, South Pacific
October 12, 1996	partial	Greenland, Iceland, Europe, North Africa
March 8–9, 1997	total	Russia, eastern Asia, Arctic, northwest North America, Japan
September 1–2, 1997	partial	Antarctica, South Pacific, New Zealand, Australia
February 26, 1998	total	Pacific, Central America, Atlantic, West Indies
August 21–22, 1998	annular	Malaysia, Indonesia, Philippines
February 16, 1999	annular	Australia
August 11, 1999	total	Atlantic, United Kingdom, Europe, India
February 5, 2000	partial	Antarctica
July 1, 2000	partial	Southern Chile
July 31, 2000	partial	Northwest Canada, Siberia, Alaska
June 21, 2001	total	Angola, Mozambique, Zambia, Madagascar
December 14, 2001	annular	Costa Rica, Nicaragua

LUNAR ECLIPSES
Lunar eclipses can be either partial (when the Moon is only partially covered by the Earth's shadow) or total (when the Moon is totally covered by the Earth's shadow).

DATE	TYPE OF ECLIPSE	AREA FROM WHICH ECLIPSE CAN BEST BE VIEWED
April 3–4, 1996	total	Africa, South America, Europe
September 27, 1996	total	Nth America, Central America, Sth America, Europe, Africa
March 24, 1997	partial	North America, Alaska, Hawaii
September 16, 1997	total	Asia, Africa, Europe, Australasia
July 28, 1999	partial	North America
January 21, 2000	total	North and South America
July 1, 2000	partial	Southern Chile
July 16, 2000	total	Pacific Ocean, Australia, eastern Asia
January 9, 2001	total	Asia, Africa, Europe
July 5, 2001	partial	Australia, eastern Asia

SELECTED MILESTONES IN SPACE EXPLORATION
Since the mid-1950s, space has been explored by many types of spacecraft. Satellites, rockets, probes (to the Moon and between the planets) and shuttles have supplied us with a wealth of knowledge about our solar system.

October 4, 1957	September 15, 1959	April 12, 1961	December 14, 1962	July 31, 1964	July 14, 1965	March 1, 1966
Sputnik 1 (USSR) First satellite launched into space.	*Luna 2* (USSR) First rocket reached the Moon.	*Vostok 1* (USSR) Yuri Gagarin first human in space.	*Mariner 2* (USA) Flew past Venus.	*Ranger 7* (USA) Close-range photographs of Moon.	*Mariner 4* (USA) Flew past Mars.	*Venera 3* (USSR) Landed on Venus.

PLANET FACTS

A comparison of the planets that make up our solar system shows the vast differences between them.

PLANET	DISTANCE FROM SUN (million miles/km)	MASS (as a fraction of Earth's mass)	DIAMETER (as a fraction of Earth's diameter)	NUMBER OF MOONS
Mercury	36 (58)	0.06	0.4	0
Venus	67 (108)	0.8	0.9	0
Earth	93 (150)	1.0	1.0	1
Mars	141 (228)	0.1	0.5	2
Jupiter	482 (778)	318	11.2	16
Saturn	885 (1,427)	95	9.4	18
Uranus	1,780 (2,871)	14.5	4.0	15
Neptune	2,788 (4,497)	17	3.9	8
Pluto	3,666 (5,913)	0.002	0.2	1

METEOR SHOWERS

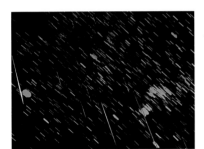

These are caused by the debris left by comets. The main annual meteor showers and their dates (which can vary by one day) are shown below. The number of meteors you see depends on the strength of the shower (sometimes as many as 50 meteors in one hour), how much moonlight is in the sky and whether you are watching them from a city or country area.

NAME OF SHOWER	DATE OF MAXIMUM ACTIVITY	COMMENT
Quadrantids	January 3	last only a few hours
Lyrids	April 22	from Comet Thatcher—produce some very bright meteors
Eta Aquarids	May 5	from Comet Halley
Delta Aquarids	July 30	a strong shower, especially with the help of the Perseids
Perseids	August 12	from Comet Swift-Tuttle
Orionids	October 22	from Comet Halley
Taurids	November 3–5	from Comet Encke—fireballs
Leonids	November 18	from Comet Tempel-Tuttle—major storm possible in 1999
Geminids	December 14	these and Perseids are the year's best showers

---

July 20, 1969
Apollo 11 (USA)
Humans landed on the Moon.

December 3, 1973
Pioneer 10 (USA)
Flew past Jupiter.

July 20, 1976
Viking 1 (USA)
Landed on Mars to search for life.

September 1, 1979
Pioneer 11 (USA)
Flew past Saturn.

January 24, 1986
Voyager 2 (USA)
Flew past Uranus.

March 6, 1986
Vega 1 (USSR)
Photographs of Comet Halley.

August 25, 1989
Voyager 2 (USA)
Flew past Neptune.

Glossary

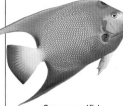
Scallop shells

Queen angelfish

Submersible

Emperor penguins

Black marlin tail

active volcano A volcano that can erupt at any time.

algae The simplest forms of plant life.

ankylosaurs Members of a group of late Cretaceous dinosaurs that spread throughout North America and East Asia. They were heavily armored with thick plates of bone, spikes, and bony nodules in the skin of their back and sides.

Antarctic The extremely cold region at the South Pole, which is south of the Antarctic Circle.

archosaurs A major group of reptiles that includes the living crocodilians as well as the extinct dinosaurs, pterosaurs, and thecodontians.

Arctic The very cold region at the North Pole, which is north of the Arctic Circle.

ash Small fragments of rock and lava blown out of a volcano during an eruption.

asteroid A small, rocky body that can be as large as 600 miles (about 1,000 km) in diameter. Most asteroids orbit in the main asteroid belt between Mars and Jupiter.

astronomy The scientific study of the solar system, our galaxy, and the universe.

atmosphere The layer of gases that surrounds a planet.

atmospheric pressure The force exerted by air on its surroundings.

atom The smallest piece that an element can be divided into and still keep its chemical properties.

aurora A display of colored lights in the sky that makes the Earth's upper atmosphere glow. It is caused by streams of particles from the Sun and happens most around the polar regions.

axis An imaginary line through the center of planets and satellites around which they rotate.

barometer An instrument for measuring atmospheric pressure.

Beaufort Scale This is used to indicate the strength of the wind at sea. It was named after Francis Beaufort, a British admiral.

Big Bang theory The theory that the universe was formed as the result of a massive expansion of matter and energy.

Big Crunch theory The theory that the universe as we know it will collapse back in on itself and end.

bipedal Walking on two legs.

black hole The last stage in the life of a galaxy or a massive star when it collapses and keeps on collapsing until even light cannot escape from it.

blizzard A snowstorm with strong winds.

caldera A crater usually more than 3 miles (5 km) in diameter formed by an explosion or powerful volcanic eruption.

carnosaurs Large carnivorous, or meat-eating, theropod (saurischian) dinosaurs such as *Megalosaurus*, *Allosaurus*, and *Tyrannosaurus*.

Cenozoic Era This period began with the extinction of the dinosaurs 65 million years ago and is known as the Age of Mammals.

ceratopians Late Cretaceous horned dinosaurs that existed for 20 million years.

ceratosaurs Medium-sized theropods distinguished by the small crests or bony horns on their noses.

climate The average weather conditions in a particular region over a period of at least 30 years.

cloud A visible mass of water droplets and ice floating in the air, formed when water condenses.

cluster A gathering of stars or galaxies bound together by gravity.

coelurosaurs Small, light, carnivorous saurischians such as *Coelurus*, *Compsognathus*, *Gallimimus*, and *Struthiomimus*.

comet A body of dust and ice that orbits the Sun. As the comet nears the Sun, it boils and becomes brighter. It forms a coma of gas and dust and sometimes one or two prominent tails.

continent One of the seven main land masses of the globe: Europe, Asia, Africa, North America, South America, Australia, Antarctica.

continental drift A theory proposed by Alfred Wegener whereby the continents were once joined together as one land mass, and then, over millions of years, drifted apart.

coral reef A structure that is made from the skeletons of soft-bodied coral animals, or polyps, and is found in warm waters.

core The core of the Earth is its central area (both solid and liquid) composed largely of iron. The core of the Sun is its inner quarter where the temperature is high enough for nuclear fusion to occur.

corona The outermost layer of the Sun's atmosphere where the temperature is nearly 2 million°F (more than 1 million°C).

crater A funnel-shaped opening at the top of a volcano. Craters usually have a diameter of 6/10 mile (1 km) or less.

Cretaceous Period Geological period from 145 to 65 million years ago. This period saw both the flowering and the extinction of the dinosaurs.

crocodilians The only living representatives from the archosaur group of reptiles. They include crocodiles, gharials, tomistoma, alligators, and caimans.

crust The hard, outer layer of the Earth, which is closest to the surface. The crust under the continents is usually about 25 miles (40 km) thick. The crust under the oceans, however, is only about 3 miles (5 km) thick.

crustacean An animal such as a lobster, crab, or prawn that has a hard skeleton on the outside of its body.

current A flow of water or air.

cyclone A violent tropical storm, also known as a hurricane or a typhoon.

deforestation Widespread clearing of a forest.

desert An area that receives little rain.

dormant volcano A volcano that is not currently active but could erupt again.

drizzle Light rain, with water drops less than 1/5 inch (0.5 mm) in diameter.

drought A prolonged period without any rain.

equator An imaginary line that lies halfway between the North and South poles.

equinox Either of the two occasions, six months apart, when day and night are of equal length.

evolution Gradual change. New species of dinosaurs evolved over millions of years.

extinct volcano A volcano unlikely to erupt again.

extinction The dying out of a species. All of the remaining dinosaur species became extinct in the Cretaceous Period.

fault A crack or break in the Earth's crust where rocks have shifted.

fog A dense, low cloud of water droplets lying near to the ground, which reduces visibility to less than 3,608 feet (1,100 m).

fossil The remains or imprint of a plant or animal found in rock.

frost An icy coating that forms when moisture in the air freezes.

galaxy A large cluster of billions of stars and clouds of gas and dust, held together by gravity.

gas giant A large planet with a very deep atmosphere and perhaps no solid surface.

geyser A hot spring that boils and erupts hot water and steam.

glacier A slow-moving mass of ice, formed in mountains, which creeps down valleys.

global warming The gradual increase in the average global temperature from year to year.

Gondwana The southern supercontinent formed when Pangaea split into two, which began about 208 million years ago. Gondwana included today's land masses of South America, Africa, India, Australia, Antarctica, and small parts of other continents.

gravity The force that attracts or draws one body to another.

greenhouse effect The increase in the temperature of a planet caused when its atmosphere prevents heat from escaping. The atmosphere surrounding the planet acts in the same way as the glass of a greenhouse.

hadrosaurs Duckbilled dinosaurs such as *Hadrosaurus*, *Maiasaura*, or *Anatotitan*. Hadrosaurs were the most common and varied plant-eating ornithopods.

hail Hard, icy pellets formed in cumulonimbus clouds, which are solid when they reach the ground.

hotspot volcano A volcano that forms in the middle of a plate, above a source of magma.

humidity The amount of water vapor in the air.

hurricane A large tropical depression with high winds and torrential rainfall. Also called a cyclone or a typhoon.

ice age A cold period during which ice extends over as much as one-third of the Earth's surface.

iceberg A large, floating chunk of ice, broken off from a glacier and carried out to sea.

Tuojiangosaurus' tail

Parasaurolophus

Teeth of plant-eating dinosaurs

Triceratops

Pleisochelys

Radiosonde

Seasons

Rainbow

Grasshopper

iguanodonts Large, plant-eating, ornithopod dinosaurs. Iguanodonts evolved in the mid-Jurassic Period and spread throughout the world.

inner core The solid ball of iron and nickel at the center of the Earth.

invertebrate An animal without a backbone.

island arc A volcanic island chain that forms when magma rises from a subduction zone.

isobar A line on a map joining points of equal atmospheric pressure.

Jurassic Period The second geological period in the Age of Reptiles. It lasted from 208 to 145 million years ago.

Laurasia A large continent, once part of Pangaea, which included today's continents of North America, Europe, and Asia and the island of Greenland.

lava Magma that erupts at the surface of the Earth.

light year The distance a beam of light travels through space in one year.

lightning A flash of electricity in the sky usually generated during a thunderstorm.

lithosphere The two layers of solid rock, consisting of the upper mantle and the crust, which lie above the asthenosphere.

magma Molten rock inside the Earth.

mantle The layer of the Earth between the crust and the outer core. The mantle has three sections: the solid lower mantle, the squishy asthenosphere, and the solid upper mantle.

mass The amount of matter in a body.

meteoroid A very small body that travels through space. If it passes through the Earth's atmosphere, it is then called a meteor.

meteorology The study of the weather.

mineral A material appearing in nature that is extracted by mining.

monsoon A wind that changes direction, bringing heavy rain during the wet season.

moon A natural satellite that orbits a planet.

navigation The science of directing the course of a ship or an aircraft.

nebula A cloudlike patch in the sky made up of gas and dust. Some nebulae are the birthplaces of stars.

ocean A very large stretch of water.

ocean current A huge mass of water that travels enormous distances and mixes warm water near the equator with cold water from the polar regions.

oceanography The science of the features and the structure of the ocean.

orbit The invisible curved path followed by one object, such as the Moon, around another object, such as the Earth.

ornithischians "Bird-hipped" dinosaurs. All the ornithischian dinosaurs were plant eaters.

outer core The liquid layer of iron and nickel surrounding the inner core.

ozone layer A diffuse layer of ozone molecules found high in the atmosphere. It filters out harmful ultraviolet radiation from the sun.

pachycephalosaurs Plant-eating, late Cretaceous, ornithopod dinosaurs with skulls thickened into domes of bone.

pack ice Ice that forms when the surface of the ocean freezes.

paleontologist A scientist who studies ancient life, especially the fossils of plants and animals.

Pangaea The supercontinent that formed in the Permian Period and broke up during the Jurassic Period. It included today's continents in one large land mass.

phases The cycle of changes, as seen from the Earth, in the shape of the Moon, Mercury, and Venus as they revolve around the Earth or Sun.

plesiosaurs Large marine reptiles of the Mesozoic Era.

pliosaurs Plesiosaurs with short necks and thick, powerful bodies.

precipitation Water or ice, such as snow, sleet, or rain, which falls to the ground from clouds.

prosauropods Late Triassic to early Jurassic ancestors of the long-necked sauropods.

pterosaurs Flying reptiles, only distantly related to dinosaurs.

quadrupedal Walking on four legs.

quasar A distant, starlike object with an enormous energy output many times brighter than an ordinary galaxy.

rainfall The amount of rain received by a particular region over a set period.

red giant A large star with a relatively cool surface.

revolution The motion of a planetary object along its orbit.

rhyolite A light-colored igneous rock formed from thick, sticky lava.

Richter scale A scale that measures the amount, or magnitude, of energy released by an earthquake. It is expressed in Arabic numbers.

rift The crack or valley that forms when two plates with ocean crust move apart.

rotation The spinning of the Earth or another body in space on its own axis.

satellite A small body in space that orbits a larger one. The Earth's Moon is a satellite.

saurischians "Lizard-hipped" dinosaurs.

sauropods Large, plant-eating saurischian dinosaurs.

sea A body of water that is partly or completely enclosed by land.

seamount A large, underwater volcanic mountain that remains under the sea or rises above it to form an island.

season A weather period of the year.

shock waves The energy that is generated by an earthquake and travels in the form of waves through the surrounding rocks.

shooting star The informal name given to a meteor when it passes through the Earth's upper atmosphere and burns up.

sleet A mixture of snow and rain.

smog A fog contaminated with air pollutants, which react together in the presence of sunlight.

snow Falling ice crystals.

solar flare A sudden, violent explosion of energy that occurs on the surface of the Sun.

solar system The collection of planets, asteroids, comets, and dust that orbits the Sun.

solstice Either the shortest day of the year (winter solstice) or the longest day of the year (summer solstice).

spectrum The rainbow colors that white light produces when it passes through a water droplet or a glass prism.

stegosaurs Late Jurassic dinosaurs that had alternating or staggered rows of plates along their backs, and two pairs of long, sharp spikes at the end of their strong tails.

subduction The process of one plate slowly diving beneath another.

supercluster A giant grouping of many clusters of galaxies. The local supercluster, for example, includes our own group of galaxies as well as many nearby, related clusters.

supergiant A very large star that has great mass and is extremely bright.

supernova A stage in the life of a massive star when the star collapses in on itself and then explodes.

synoptic chart A weather map showing conditions at a particular point in time.

temperate Moderate or mild weather conditions.

temperature A measure of the amount of heat.

thermometer An instrument that measures the temperature.

theropods Meat-eating saurischian dinosaurs that walked on their hind legs.

thunder A rumbling shock wave created when lightning heats the air.

tide The repeating rise and fall of the Earth's seas, caused by the pull of the moon and sun on the water.

tornado A violent, spiraling wind that is short-lived but destructive.

Triassic Period The first period of the Age of Reptiles. The Triassic Period lasted from 245 to 208 million years ago.

tropical Describing hot and often humid conditions experienced in regions close to the equator.

tsunami A gigantic, often destructive, sea wave which is triggered by an underwater earthquake or volcano.

typhoon A violent tropical storm, also known as a hurricane or cyclone.

vertebrae Bones from the back of the skull to the tail that protect the spinal column.

volcanic islands Underwater volcanoes are called seamounts. Some may break the surface of the ocean and form volcanic islands.

volcanologist A scientist who studies volcanoes.

weather The atmospheric conditions experienced at a particular place or time.

wind A mass of air that moves from one place to another.

Pluto's strange orbit

Newton's telescope

Uranus

Mars Rover

Daytime moon

221

Index

abyssal plains, 57
Adams, John, 193
adaptation, 72
advection fog, 147
African plate, 112
aftershocks, 124
air currents, 139, 142
Albertosaurus, 21
algae, 8, 9, 32, 87
Allosaurus, 13, 16, 17, 20, 50
Alphadon, 19, 46
altocumulus clouds, 140, 141
altostratus clouds, 140, 141
Alvin, 72, 74, 75
ammonia, 186, 192
Anak Krakatau, 115
Anatotitan, 31
anchovies, 68, 89
Andromeda Galaxy, 202, 208
anemometers, 150, 151
anemones, 62, 66
anglerfish, 71
animals
 climatic change and, 168-9, 170
 desert zone, 166, 167
 mountain zone, 160, 161
 polar zone, 159
 temperate zone, 162
 tropical zone, 164
Ankylosaurus, 26
Antarctica, 137, 158, 168
Apatosaurus, 10-11, 22, 44
Aphrodite, 179
Apollo, 203, 212, 217
Archaeopteryx, 17, 48, 49
archosaurs, 48
Arctic, 158, 159
Arctic terns, 82
Arenal, 127
armored dinosaurs, 22, 26-7, 36, 44
armor-plated fish, 8
Armstrong, Neil, 182, 212
ash and gas, 106-7, 111, 113, 114, 117, 119
asteroid belt, 198
asteroids, 174, 179, 198, 199
asthenosphere, 92, 94, 97, 98, 99
Atlantic Ocean, 96
Atlantis, 80, 81
atmosphere, 132, 134, 135, 150
 Earth, 180
 Mars, 184
 Neptune, 192
 Pluto, 194
 Titan, 189
atmospheric pressure, 132, 134
auroras, 177
axis, 174

bacteria 8, 9, 32, 70, 71
Bactrosaurus, 19
ball lightning, 142
balloons, 150
Baryonyx, 21
basalt, 104, 110, 111
bathythermographs, 56
Beaufort Scale, 59
Bermuda Triangle, 80, 81
Big Bang theory, 202, 203
Big Crunch theory, 202, 203
binary star system, 206-7
bioluminescence, 70, 71
bipedal dinosaurs, 12, 16, 34
bird-hipped dinosaurs, 12, 13, 16, 22
bird-mimic dinosaurs, 44
birds, 8, 13, 47, 48, 49, 82, 87
black holes, 205, 209, 210
black smokers, 73, 97

blizzards, 158, 159
blue giants, 204
bluefish, 64
body temperature, 40
bones *see* fossils
Brachiosaurus, 17, 22, 24, 25, 30, 35, 41, 51
brain size, 20, 35
"Breadknife", 110
Brocken Specters, 148
buildings, earthquake-proof, 120, 121

calderas, 108-9, 115
California, 128-9
Callisto (moon), 186, 187
Camarasaurus, 23, 34
camouflage, 44, 70
Campbell-Stokes recorder, 150
Camptosaurus, 17, 32, 34
Cappadoccia, 110-11
carbon, 202
carbon dioxide, 170, 171, 178
Caribbean plate, 94, 126, 127
Carina, 211
Carnegie, Andrew, 25
carnosaurs, 15, 20, 21
Cassini's division, 188
Celsius, Anders, 137
Centrosaurus, 26
Cepheid variables, 206
ceratopians, 26
ceratosaurs, 20
Ceratosaurus, 35
Ceres, 198
Challenger program, 199
Charon (moon), 194, 195
Chasmosaurus, 26
chitons, 63
chlorofluorocarbons (CFCs), 171
cirrocumulus clouds, 140
cirrostratus clouds, 140
cirrus clouds, 140
clams, 62, 66, 67, 72
climate, 60, 134, 135, 154, 156-67
climatic changes, 56-9, 163
cloud symbols, 152
clouds, 135, 138-43, 144, 166
coal, 16
coastal seas, 64-5
Cocos plate, 94, 126, 127
Coelophysis, 14, 15, 50
coelurosaurs, 15, 20, 21
Coelurus, 17, 32, 50
cold temperate regions, 162
colored rain, 148
colors and markings, 37
coma, 196
Comet Encke, 217
Comet Levy, 197
comets, 174, 179, 196-7, 186, 217
Compsognathus, 20, 21, 30, 48, 49
computers, 152, 153, 154
condensation, 138, 139, 146, 147, 156, 164
conservation, 88-9
constellations, 204
contact binaries, 206-7
continental crust, 92, 98, 99
continental drift, 94
continental shelf, 56
continental slope, 56
convection currents, 92, 94-5, 139, 142
cool temperate regions, 162, 163
coral reefs, 62, 66-7, 88
Coriolis effect, 61, 155
corona, 176, 200
Corythosaurus, 19, 23, 35
Cousteau, Jacques, 75

crabs, 62, 86
crater lakes, 108, 109
craters, 103, 108-9, 174, 178, 179, 182, 184, 187, 188, 199
crescent phase (Moon), 182
Cretaceous Period, 10, 11, 18-19, 22, 32, 46, 50
crocodiles, 8, 14, 47, 48
crust, 92, 94
crustaceans, 62, 67, 68, 70
crystals, 138, 143, 144, 145, 146, 148
cumulonimbus clouds, 140, 141, 142, 144
cumulus clouds, 139, 140, 141, 142
currents, 60-1
 air, 135, 138, 139, 142, 143, 144, 145, 161
 ocean, 154-5, 156
cycads, 17, 22

daylight, 180, 186, 190
deep-sea fish, 70-1
deforestation, 170, 171
Deimos (moon), 184
Deinonychus, 20, 35, 45, 51
Delta Aquarids, 217
deserts, 136, 137, 156, 157, 166-7
Dimetrodon, 8, 11
Diplodocus, 16, 22, 24-5, 31, 44
Disney, Walt, 195
diving, 56, 74-5, 85
doldrums, 154, 155
dormant volcanoes, 100, 102, 108
dragonfish, 71
Dromaeosaurus, 19
Dromiceiomimus, 41
droughts, 132, 133
duckbilled dinosaurs, 18, 19, 22, 28-9, 37, 41-2, 43
dung, fossilized, 32, 36, 41-2, 43, 62
Dunkleosteus, 8
dwarf stars, 204, 206

Earth, 60, 61, 174, 180-1, 193, 200, 201, 217
earthquake lizard, 24, 30
East Pacific Rise, 96
echolocation, 65
eclipses, 200-1, 216
Edmontosaurus, 12, 29, 41-2
eggs, 36, 38-9, 64, 67, 83
Eldfell, 116-17
elliptical galaxies, 208, 209
emergency services, 125, 126-7, 129
Encke gap, 189
Eoraptor, 8, 14
equator, 135, 156, 164
equinox, 156
Eta Aquarids, 217
Euoplocephalus, 19, 27, 44, 50
Euparkeria, 8
Eurasian plate, 112, 114, 116, 125
Europa (moon), 186, 187
evaporation, 135, 136, 139, 150
evolution, 8-9, 12, 19, 46, 50
exploding stars, 207
exploitation, 74-5, 84-7
exploration, 56, 34-77, 84-5
extinction, 46-7, 50
extrusive rock, 110
Exxon Valdez, 87

Fahrenheit, Gabriel, 137
faults, 95
Ferrel cells, 155
fires, 124, 129
fish, 62, 64, 65-71
fishing, 88

floods, 118, 122-3, 133
flowering plants, 18, 19, 22
fog, 146, 147, 152
food chain, 68-9, 86
food production, 133
footprints, 32, 36
forests, 14, 16, 18, 47
fossil fuel, 84-5, 170
fossil sites, 14, 16, 18, 33
fossils, 13, 14, 16, 18, 20, 31, 32-3, 48, 168
freezing, global, 168-9
fronts, 138, 152
frost, 146-7
full phase (Moon), 182

Gagarin, Yuri, 212, 216
galaxies, 202, 208-9, 210-11
Galileo, 182, 186
Gallimimus, 44
Ganymede (moon), 186, 187
gas and ash, 106-7, 111, 113, 114, 117, 119
gases, 135, 170, 174, 176, 180, 186, 187, 202
Geminids, 217
geothermal energy, 116
geysers, 101
gibbous phase (Moon), 182
glaciers, 168
global freezing, 168-9
global warming, 170-1
Glomar Challenge, 56
Gondwana, 16, 18, 94
granite, 110, 111
gravity, 174, 182, 186, 188, 206, 207, 209, 213
Great Barrier Reef, 57, 88, 89
Great Dark Spot, 192
Great Kanto earthquake, 124-5
Great Nebula, 204
Great Red Spot, 187
greenhouse effect, 170, 178
ground fog, 147
Gulf Stream, 60, 154, 156
guyots, 57
gyres, 61, 154

habitat loss, 46, 47
Hadley cells, 155
hadrosaurs, 28, 36
hail, 144, 145
Halley's Comet, 197, 217
Hawaii, 100, 101
Hawaiian eruptions, 102, 104-5
helium, 176, 188, 190, 202
Herculaneum, 112-13
herds, 22, 24, 26, 36, 44
Herrerasaurus, 15
herrings, 64, 69, 70
Herschel (crater), 188
Herschel, William, 190, 191
Heterodontosaurus, 23
Himalayan mountain system, 135, 160
hips, 12-13
 ornithischian, 12
 saurischian, 13
horned dinosaurs, 22, 26-7, 36, 41-2
Horner, Dr. John, 38
Horsehead Nebula, 205
hot springs, 97, 101, 110
hotspot volcanoes, 100-1
humidity, 136-7, 164
humpback whales, 64-5
hurricanes, 58, 59, 132
Huygens, Christiaan, 188
hydrogen, 176, 186, 188, 190, 192, 202, 204, 205, 206, 207

Hydrolab, 74
hydrothermal vents, 97
hygrometers, 136, 137, 150
Hylonomus, 8
Hypacrosaurus, 19
Hypsilophodon, 34, 51

ice, 135, 138, 143, 144, 145, 146–7, 158, 160, 168
ice ages, 168
Iceland, 96, 116–17
Ichthyostega, 8
igneous rock, 110
Iguanodon, 13, 23, 35, 44
Indo-Australian plate, 114
Indonesia, 103, 109, 111, 114–15, 122
inner core, 92, 180
interglacial periods, 168
intrusive rock, 110
Io (moon), 186, 187, 193
irregular galaxies, 208, 209
Ishtar, 179
island arcs, 98, 99
islands, 96, 97, 100
isobars, 152
Italy, 112–13

Japan, 121, 122, 124–5
Jason Jr., 74
jaws, 20, 34
jellyfish, 8, 9
Jensen, James, 31
jet streams, 154, 155
Juan de Fuca plate, 118
Jupiter, 174, 186–7, 188, 217
Jurassic Period, 10, 11, 16–17, 22, 32, 50

Kawah Ijen, 111
Keli Mutu, 109
Kentrosaurus, 27
Kilauea, 100, 101, 108
kimberlite, 111
Krafft, Maurice and Katia, 107
Krafla fissure, 117
Krakatau, 114–15
krill, 68, 69, 82
Kronosaurus, 11

Lagosuchus, 8
Lakagigar fissure, 117
Lambeosaurus, 37
landforms and rocks, volcanic, 108, 110–11
Large Magellanic Cloud, 208
larvae, 68, 83
Laurasia, 16, 18, 94
lava, 92–3, 100, 102, 104–5, 116–17
lava lakes, 92, 108
Le Verrier, Urbain, 193
Leonids, 217
life cycle, 19
life forms, 62–73, 82–3, 89
light, 70, 71, 72, 176, 208
light years, 202
lightning, 142–3
limestone, 66, 67
liquefaction, 121
lithosphere, 92, 94, 95, 97, 98, 99, 101, 128
lithospheric plate, 94
lizard-hipped dinosaurs, 12, 16, 22
lizards, 8, 14
lobsters, 83, 89
Local Group, 208
Loch Ness Monster, 79
Loihi seamount, 101
Loma Prieta, 128, 129
long-necked dinosaurs, 22, 24–5, 31
Los Angeles, 128, 129
Lowell, Percival, 184, 195
Luna 2, 216
lunar eclipses, 200–1, 216
Lyrids, 217

Magellan, Ferdinand, 209
Magellanic Clouds, 208, 209
magma, 92, 94, 96–8, 100, 101–3, 106, 108, 110, 116
magma chambers, 97, 103, 109
magma plume, 101
Maiasaura, 38, 39, 50
Mamenchisaurus, 24–5, 31
mammatus clouds, 140
mammoths, 168–9
mantle, 92, 95–8, 180
marine parks, 88
Mariner program, 179, 185, 216
Mars, 174, 184–5, 212, 217
Mars Rover, 184
Mary Celeste, 80
Mauna Loa, 100, 101
meat-eating dinosaurs 12, 15, 20–1, 34, 41–2, 44, 50–1
Megalosaurus, 20, 30
Mercury, 174, 178–9, 217
mermaids/mermen, 78, 79
mesosphere, 135
Mesozoic Era, 10, 111
meteorites, 46, 199
meteoroids, 198
meteorology, 151, 152
meteors, 198, 217
methane, 170, 186, 189, 192
Mexico, 126–7
midnight sun, 180
mid-ocean ridges, 92, 95, 98
migration, 82–3
Milky Way, 202, 208, 210–11
Mimas (moon), 188
minerals and oils, 84–5
Mir space station, 212
Miranda (moon), 190
mollusks, 62, 67
molten magma, 110
monsoons, 135, 164
monsters, 78–9
moons
 Earth, 60, 180, 182–3, 200–1, 212, 213, 217
 Jupiter, 180, 186, 187, 217
 Mars, 184, 217
 Neptune, 192, 193, 217
 Pluto, 194, 217
 Saturn, 188, 189, 217
 Uranus, 190, 217
mosasaurs, 47
Mt. Agung, 103
Mt. Erciyes, 110–11
Mt. Etna, 112
Mt. Ngauruhoe, 108
Mt. St. Helens, 118–19
Mt. Vesuvius, 112
mountain zones, 135, 139, 156, 157, 160–1
mountains, 56, 57, 94, 95, 99
mudflows, 118, 119

nautilus, 64
Nazca plate, 94, 126, 127
neap tides, 60
nebula, 204, 205
Neptune, 78, 79, 174, 175, 188, 192–3, 195, 217
nests, 36, 38–9
New Guinea, 99
New Zealand, 108
Nicaragua, 123
nimbostratus clouds, 140, 144
nitrogen, 189, 192, 193
North American plate, 100, 116, 126, 128
Northern Hemisphere, 60
northern lights, 177
nuclear fusion, 176, 207
Nyiragongo, 92–3

obsidian, 104, 110
ocean/sea floor, 56–61, 72–3
ocean crust, 92, 94, 96, 98, 99

ocean floor spreading, 96, 97, 98
ocean trenches, 94, 98, 99
oceans, 135, 147, 154–5, 156
octopuses, 63
oil spills, 86, 87
oils and minerals, 84–5
Olympus Mons, 184
Orion, 204, 211
Orionids, 217
Ornitholestes, 10–11
Ornithosuchus, 8
Ostro, Steve, 199
Othnielia, 22
Ouranosaurus, 35, 40–1, 51
otters, sea, 63
Owen, Sir Richard, 11, 51
Oviraptor, 21, 38, 39
oxygen, 180
ozone layer, 135, 171

pachycephalosaurs, 22, 26
Pachycephalosaurus, 19, 44, 45, 50
Pacific plate, 99, 100, 125, 126, 128
Pacific Ring of Fire, 99, 127
pack ice, 158
Paleozoic Era, 8, 9, 11
Pangaea, 14, 16, 94
Papua New Guinea, 99
Parasaurolophus, 28, 29, 51
Peléean eruptions, 102, 108
Penzias, Arno, 203
Perseids, 217
Philippine plate, 94, 125
Philippines, 106
Phobos (moon), 184
photosphere, 176
photosynthesis, 66, 170
phytoplankton, 68
Pioneer program, 217
Piton de la Fournaise, 100
plains, 56, 57
plankton, 64, 68, 69, 89
Planocephalosaurus, 14
plant-eating dinosaurs, 15, 18, 22–3, 26, 28, 41–2, 44, 50–1
plants
 climatic change and, 168, 170, 171
 desert zone, 166, 167
 mountain zone, 160
 temperate zone, 162, 163
 tropical zone, 164, 165
plated dinosaurs, 26–7, 36, 41
Plateosaurus, 15, 23, 50
plates, 94–5, 96, 97, 98, 99, 100, 126, 127, 128, 129
Pleisochelys, 17
plesiosaurs, 47
Plinian eruptions, 102, 108, 119
pliosaurs, 47
Pluto, 174, 175, 194–5, 217
pluviographs, 150
Polacanthus, 27
polar winds, 155
polar zones, 134, 135, 137, 157, 158–9, 168
pollution, 86–9, 146, 170
polyps, 66, 67
Pompeii, 112
Portuguese man o' war, 68
Precambrian Era, 9
precipitation, 144, 160
Prenocephale, 18
pressure, atmospheric, 132, 134
prevailing winds, 154, 155
Procompsognathus, 15
Project Phoenix, 214
prominences, 176
prosauropods, 15
Protoceratops, 18, 23
protostars, 204, 205
Proxima Centauri, 204
pterosaurs, 11, 46, 47
Ptolemy, 155

pulsars, 206
pumice, 106, 107, 110, 114, 115
pyroclastic flows, 107

Quadrantids, 217
quadrupedal dinosaurs, 12, 16, 34
quasars, 209

radar domes, 151
radiation, 203
radiation fog, 147
rain gauges, 150
rainbows, 133, 148–9
rainfall, 133, 135, 139, 143, 144, 148, 156, 164, 166
rainforests, 134, 136, 164
Ranger 216
razor clams, 62
red giants, 204, 205, 206
reflection, 134, 135, 138, 148, 149, 158
reforestation, 171
reptiles, 8, 9, 14, 46
rhyolite, 104, 110
ribbon lightning, 142
Richter scale, 124, 126
ridges, 56, 57, 95, 96–7, 98
rifts, 96–7, 116
rings, planetary, 186, 188–9, 190–1, 192
rocks and landforms, volcanic, 108, 110–11
rotation, 174, 175, 180, 181, 182, 186, 188

salmon, 68, 69, 70, 83, 89
Saltasaurus, 19, 22, 50
Saltopus, 15
San Andreas Fault, 128, 129
San Francisco, 128–9
sand dunes, 166–7
sardines, 64, 68, 89
satellite dishes, 151
satellites, 151, 152, 212, 216
Saturn, 174, 175, 188–9, 217
Saurolophus, 29
sauropods, 12, 16, 22, 24, 34, 36, 38, 41, 43, 44
Scaphognathus, 11
Schiaparelli, G., 184, 185
science fiction, 185, 214–15
scuba divers, 74, 85
sea floor spreading, 96, 97, 98
sea lions, 68, 69
sea urchins, 62, 63, 67
seals, 68, 69, 86–7
seashore, 62–3
seasons, 156, 158, 162–3, 164, 165, 180, 184
secondary waves, 124
Seismosaurus, 24, 30, 31
shellfish, 62, 64
ships, 76–7
 weather, 151, 152
shock waves, 126
shooting stars, 198
silica, 104
sills, 103, 110
Sirius, 176
skeletons, 34–5, 35–8
skin, 36, 37
skulls, 20, 22, 34–5, 48
sleet, 144, 152
Sloane's viper, 70
Small Magellanic Cloud, 208
smog, 146
smoke, 102–3
smoking chimneys, 96
snakes, sea, 64–5
snow, 132, 134, 135, 144, 145, 152, 158, 159, 160, 161, 162, 163
snowflakes, 145
solar eclipses, 200, 201, 216
solar energy, 134–5, 139, 156, 158, 166
solar flares, 176, 177

solar prominences, 176
solar system, 174-5, 208
solstices, 156
sound, 57, 65
South American plate, 126, 127
South Bismarck plate, 99
Southern Hemisphere, 60
southern lights, 177
space exploration, 184, 186, 188, 190, 199, 212-13, 216-17
space stations, 212
spacemobiles, 184-5
spectrum, 149
spiral galaxies, 208, 209
spreading ridges, 96, 97
spring tides, 60
Sputnik 1, 212, 216
squid, 68, 69, 70, 71
star trails, 181
Star Trek, 214, 215
starfish, 62
stars, 204-11
steam, 101, 102-3
stegosaurs, 26
Stegosaurus, 16, 17, 26, 27, 50
Stevenson screens, 150
stomach stones, 22, 41-2, 43
storms, 59, 132, 138, 142, 143, 144, 145, 166
stratocumulus clouds, 140
stratosphere, 135
stratus clouds, 140, 141, 144
Strombolian eruptions, 102
Struthiomimus, 30, 51
Styracosaurus, 26, 41-2
subduction, 92, 98-9, 100, 104, 125
submersibles, 72, 74-5

subtropical regions, 164, 165
sulfur, 96, 111, 187
sun, 60, 66, 68, 134-5, 139, 156, 158, 166, 174, 176-7, 178, 180, 190, 200-1, 204, 211
sundogs, 148
sunspots, 176, 177
superclusters, 202, 208
supernovas, 205
Supersaurus, 31
surgeonfish, 66
Surtsey Island, 97
surviving an earthquake, 120-1
sweating, 136
synoptic charts, 152-3

tails, 25, 35, 44
Tambora, 114
Taurids, 217
teeth, 20, 22, 23, 29, 31, 35, 36, 41-2, 44
telescopes, 191
temperate zones, 157, 162-3
temperature, 72, 132, 134, 135, 136-7, 146, 154, 157-60, 162, 164, 165
Tenma, 212
Tenontosaurus, 20, 45
thermometers, 150
thermosphere, 135
theropods, 16, 20
thunder, 142-4
thunderclouds, 141-3
tides, 60-1, 182
tilt, 190, 191
Titan (moon), 189
Titanic, 74
Toutatis, 199

Toyko, 121
trackways, 36
trade winds, 154, 155
tree rings, 168
trenches, 56, 57, 94, 98, 99
Triassic Period, 8, 10, 11, 14-15, 22, 50
Triceratops, 18, 19, 44, 51
trilobites, 8
tripod fish, 73
Triton (moon), 192, 193
Trojan asteroids, 199
Troodon, 38, 41-2
tropical zones, 134, 157, 164-5
troposphere, 135
tsunamis, 58, 59, 113, 114, 122-4, 126
tubeworms, 72-3
tuna, 64, 70, 89
Tuojiangosaurus, 41, 44
Turkey, 110-11
turtles, 8, 14, 17, 47, 82
twilight zone, 70-1
Tyrannosaurus, 18, 19, 20, 26, 31, 35, 41-2, 44, 51

Ultrasaurus, 31
ultraviolet light, 135, 171
Ulysses, 217
Unidentified Flying Objects, 214-15
universe, 202-3
Unzen, 97
upheavals, sea, 58-9
upslope fog, 147
Uranus, 174, 175, 188, 190-1, 217

valleys, 56, 147, 168
vapor, 135, 136, 137, 138, 139, 141
variable stars, 206

Vega 1, 217
Velociraptor, 18, 26, 27
Venera 3, 216
vents, 73
Venus, 174, 178-9, 212, 213, 217
Verne, Jules, 214
Viking 1, 217
viperfish, 70, 71
Vostok 1, 216
Voyager program, 186, 188, 192, 217
Vulcanian eruptions, 102

walruses, 65
warm temperate regions, 162, 163
water cycle, 139
waterspouts, 59
waves, 58, 59
weather satellites, 151, 152
weather stations, 150-2
weevertish, 62
Wegener, Alfred, 94
whales, 64-5, 68, 89
whelks, 63, 82
whirlpools, 58
white dwarfs, 204, 206
Wilson, Robert, 203
wind, 58, 59, 60, 132, 134, 154-6, 159, 165, 166
worms, 72-3

Yellowstone National Park, 101
yellowtail, 64

Zaire, 92-3
Zanclodon, 15
zooplankton, 68

Picture Credits

(t=top, b=bottom, l=left, r=right, c=center, Bg=background) **Ad-Libitum**, 28cl, 39t (S. Bowey), 61cr, 76tl, 84i, 86i, 88i, 89i, (S. Bowey), 38br (S. Bowey/Australian National Maritime Museum). **Bryan and Cherry Alexander**, 158tl, 159c, 159tr. **American Museum of Natural History**, 35cr (Neg. No. 35423/ A. E. Anderson). **Anglo-Australian Observatory**, 205br, 205cr, 209r (D. Malin). **Ann Ronan Picture Library**, 149br. **Ardea**, 13tc (M. D. England), 36cl (F. Gohier). **AT & T Archives**, 203br. **Auscape**, 49br (Ferrero/Labat), 88bl (K. Deacon), 82bl (J. P. Ferrero), 89br (F. Gohier), 63cl (C. A. Henley), 64bc, 67tl, (D. Parer & E. Parer-Cook), 67cr (A. Ziebell), 107tc (Explorer/K. Krafft), 101tr (J. Foott), 109fr (F. Gohier), 92-93c, 100br, 101tl, 119tr (M. Krafft), 110br (W. Lawler, 167tl (J.M. La Roque), 160br (M. Newman). **David Austen**, 111tr. **Austral International**, 36tc (Keystone), 74cl (R. Parry/Rex Features), 129cxs, 110tl (Colorific!), 128cl (IME–Sipa-Press/ K. Levine), 129tr (Sipa-Press), 129br (Sygma/J. P. Forden), 129bcr (Sygma/L. Francis Jr, The Fresno Bee), 129cr (Topham Picture Library), 121tr (Nik Wheeler), 135cr (FPG International), 182bl (Camera Press), 195tr (Fotos International), 213cr (FPG), 184cl, 187tr (Rex Features), 176c (Rex Features/NASA), 203tl (Sipa Press/ F. Zullo). **Australian Museum**, 42c (C. Bento), 54i, 56i, 58i, 60i, 62i, 64i, 66i, 68i, 70i, 72i, 78i, 80i, 82i (H. Pinelli), 111tl (J. Fields). **Australian Picture Library**, 46cl (NASA/Reuters), 38cr (UPI/Bettmann), 65cr (Volvox), 112cl (Agence Vandystadt/ G. Planchenault), 128/129b (A. Bartel), 106br, 122tl (Reuters/Bettman), 120bl, 135bcr (R. Bisson), 133tl, 137bl, 157tl (J. Carnemolla), 145br, 161tr (ZEFA), 199br (J. Blair), 179bl (Orion Press), 213br (UPPA), 204bl. **Johnny C. Autery**, 143c. **Esther Beaton**, 86tr, 87bc, 87cl, 87tl. **Berlin Museum für Naturkunde**, 48tc (P. Wellnhofer). **Black Star**, 115tr (J. Delay), 118/119c (J. Mason), 118bl (R. Perry). **Bilderberg**, 153tr (P. Ginter). **Biofotos**, 83bl (H. Angel). **The Bridgeman Art Library**, 112bl (Phillips, The International Fine Art Auctioneers). **Bruce Coleman Ltd**, 86bl (A. Compost), 69bl (I. Everson), 67br, 67cl (C. & S. Hood), 83c (Jeff Foott Productions), 66–67cr, 67tr (J. Murray), 63cr, 63tl (F. Sauer), 66tl (N. Sefton), 86bl (J. Topham), 66bc (B. Wood), 110bl (S. Kaufman), 168tc (E. Pott), 157tcr (A. Price), 144bl (H. Reinhard), 159tl (J. Shaw), 148bl (U. Walz). **Bruce Coleman, Inc**, 119cr (J. Balog). **Christine Osborne Pictures**, 157tr (C. Barton). **Coo-ee Picture Library**, 8cl (R. Ryan). **Earth Images**, 119br (B. Thompson), 146tl (T. Domico), 140c (A. Ruid). **Ecoscene**, 160cl (Chelmick). **Fairfax Photo Library**, 99tr, (R. Stevens). **Frank Lane Picture Archive**, 145tr (J.C. Allen), 142b (R. Jennings). **Fraser Goff**, 118cl (J. Hughes). **Akira Fujii**, 179br, 182t. **The Granger Collection**, 74i, 76i, 77i, 79bcr, 81bc, 94bl, 112c, 114br, 124tr, 125tl, 137tl, 169b, 185br, 191tl, 191br, 193bc, 196tr, 197c, 215cc, 215tr. **Greenpeace**, 87br (Beltra), 87bl (Midgley). **Giraudon**, 192c (Lauros). **Richard Herrmann**, 63br. **Horizon Photo Library**, 133tr (H. Ecker), 163tl, 171tr. **Icelandic Photo and Press Service**, 117tc (S. Jonasson), 97br (G. Palsson), 116tr, 117tr (M. Wibe Lund). **The Image Bank**, 44tr (J. Hunter), 46tr (Image Makers), 104cl (L. J. Pierrce), 164tl (J. H. Carmichael), 157tc (H. G. Ross/Stockphotos, Inc), 133cl (S. Wilkinson). **Images Unlimited Inc**, 74tr (A. Giddings), 56tl (C. Nicklin). **International Photographic Library**, 170bl. **International Stock**, 142tr (B. Firth), 146c (M. P. Manheim). **J. Allan Cash Ltd**, 147tr, 148tl. **James Cook University**, 57tr (D. Johnson). **Richard Keen**, 147cr. **Franz X. Kohlhauf**, 182-183. **Landform Slides**, 109c. **Bob Litchfield**, 163tr. **Lochman Transparencies**, 141br (B. Downs). **David Levy**, 195br. **Mantis Wildlife Films**, 167c (D. Clyne). **Mary Evans Picture Library**, 78tl, 79tr, 80tl, 81br, 155br. **Matrix**, 48tl (Humboldt Museum, Berlin/L. Psihoyos), 20bl (The Natural History Museum, London/L. Psihoyos), 31cr, 43tr, 51tr (L. Psihoyos). **Mary Evans Picture Library**, 124bl. **Minden Pictures**, 86-87b (F. Lanting), 62cl (F. Nicklin), 104tl (F. Lanting), 171cr (F. Lanting). **Mirror Syndication International**, 126cl (British Museum/PCC/ Aldus). **Mt Stromlo and Siding Spring Observatories**, Australian National University, 211bl. **National Archaeological Museum of Athens**, 113br. **The Natural History Museum, London**, 13tr, 22c, 22tl, 24-25b, 25c, 36cl, 36tl, 38bl, 42bl. **National Meteorological Library**, 151bl (British Crown/By permission of the Controller of HMSO), 148tr (F. Gibson). **Paul Nevin**, 152cl. **Newell Colour**, 184bl. **NHPA**, 42tcl (P. Johnson), 70tl (Agence Nature), 79bc, 83tl (G.I. Bernard), 88cl (B. Jones & M. Shimlock), 67cr (B. Wood), 112bc (A. Nardi). **North Wind Picture Archives**, 155bl. **Oxford Scientific Films**, 63bc (A. Atkinson), 67tl (G. Could), 62tc, 66cl (H. Hall), 66tl (M. Hall), 68bc, 70tr (P. Parks), 68bl, 68cr (H. Taylor Abipp), 140bc (D. Allan), 167tr (E. Bartov), 161tl (B. Bennett). **Pacific Stock**, 108br (J. Carini). **The Photo Library, Sydney**, 33tc (SPL/S. Stammers), 84tl (H. Frieder Michler/SPL), 81bcr (D. Hardy/SPL), 69cr (NASA/SPL), 60tl (J. Sanford/SPL), 54cr, 59bl, 59cl (SPL), 87cr (V. Vick), 127tr (G. Dimijian), 103tr (A. Evrard), 125tr (Sipa Press), 108tl (R. Smith), 105bcr (TSI/G. B. Lewis), 132cr (K. Biggs/TSW), 132bl (F. Grant), 154c (Los Alamos National Laboratory/SPL), 132-133b (SPL), 135br (SPL/NASA), 181br (B. Belknap), 180b (A. Husmo), 199bl (R. Smith), 185tr, 197bc, 208br (SPL), 176bl (SPL/J. Baum), 206bl, 207br (SPL/C. Butler), 177br (SPL/J. Finch), 197br (SPL/G. Garradd), 208/209t (SPL/T. Hallas), 197tl, 214bl (SPL/D. Hardy), 217cl (SPL/D. Milon), 179tr, 186tl, 189cr, 192bl, 199cr (SPL/NASA), 200/201c (SPL/ G. Post), 176tr, 200cl (SPL/R. Royer). **Photri**, 151tr. **Photo Researchers, Inc**, 56bl (J. R. Factor), 59tl (Mary Evans), 54br (NASA/SPL), 85tr (G. Whiteley), 126/127cr (F. Gohier), 136br (D. McIntyre). **Planet Earth Pictures**, 48cl (P. Chapman), 42tl (W. Dennis), 66c (G. Bell), 87c (M. Conlin), 70cr, 71cr, 71tc, 71tr, (P. David), 56br (R. Hessler), 86l (C. Howes), 71c (K. Lucas), 62tl (J. Lythgoe), 71tcr (L. Madin), 74tl (D. Perrine), 56c (F. Schulke), 66bl (P. Scoones), 141c (J. R. Bracegirdle), 162br (J. Eastcott/Y. Momatiuk), 168br (K. Lucas), 171br (J. Lythgoe), 154tr (R. Matthews), 158/159c (K. Puttock), 182c, 182cl (R. Chesher). **Richard Wilson Studios**, 62-63c, 66-67c, 86-87c. **Robert Harding Picture Library**, 105br (A. C. Waltham), 133cr (M. Jenner), 165tl (Raleigh International), 133br, 151cr, 168cr. **Roger-Viollet**, 125bc, 125cr, 125cl (Collection VIOLLET). **Scripps Institution of Oceanography**, University of California, San Diego, 77c. **John S. Shelton**, 95br. **Marty Snyderman**, 85bl. **Sporting Pix**, 124tl (Popperfoto). **Stock Photos**, 49bcr (Animals Animals), 157tcl (H. P. Merten/TSM), 132tr (G. Monro), 139bl (R. Richardson), 138br (J. Towers/TSM), 182cr (J. P. Endress), 176br (Phototake). **Tom Stack & Associates**, 13tl (J. Cancalosi), 87cr (NASA), 190bc, 190bl (NASA/JPL), 187tc (USGS). **Topham Picturepoint**, 105cr. **University of Chicago Hospitals**, 14bl. **Woods Hole Oceanographic Institution**, 76br, 77tl. **Norbert Wu**, 71bc, 71bcr, 71br, 87tr.